INTEGRATED POLI
PREVENTION AND (

Integrated Pollution Prevention and Control

The EC Directive from a Comparative Legal and Economic Perspective

Chris Backes & Gerrit Betlem (Eds.)

KLUWER LAW INTERNATIONAL
THE HAGUE–LONDON–BOSTON

Published by
Kluwer Law International Ltd
Sterling House
66 Wilton Road
London SW1V 1DE
United Kingdom

Kluwer Law International incorporates
the publishing programmes of
Graham & Trotman Ltd,
Kluwer Law & Taxation Publishers
and Martinus Nijhoff Publishers

Sold and distributed in
the USA and Canada by
Kluwer Law International
675 Massachusetts Avenue
Cambridge MA 02139
USA

In all other countries, sold and distributed by
Kluwer Law International
P.O. Box 322
3300 AH Dordrecht
The Netherlands

ISBN 90-411-9718-4

First published 1999

British Library Cataloguing Publication Data
A catalogue record for this book is available from the British Library

Layout/editing: G.J. Wiarda Institute, Utrecht University Institute for Legal Studies, Boothstraat 6, 3512 BW Utrecht, the Netherlands.

Index typeset in 9/10 Baskerville by Creative Associates, Oxford
Printed and bound in by Athenaeum Press Ltd., Tyne and Wear

Foreword

This book is based on papers presented at a conference on the EC Council Directive 96/61/EC of 24 September 1996 concerning integrated pollution prevention and control and its consequences for the EC Member States' legal systems, held in Utrecht (the Netherlands) on 5 and 6 October 1995. The papers were prepared on the basis of the text of the Common Position of the EU Council as formally adopted on 27 November 1995, OJ 1996 C 87/8. The contributions to this book were completed at a later stage and have been updated, where appropriate.

This collection of expert papers is the first comparative commentary on the Directive which is to be implemented in EC Member States before 31 October 1999. They intend to contribute to the debate about the implementation of the Directive from a comparative legal and an economic point of view. The discussion focuses on the interaction between Community law and national law, with particular regard to Germany, the Netherlands and the United Kingdom, in order to compare those Member States with and those without previous experience of establishing a legislative and regulatory framework for integrated pollution prevention and control. For ease of reference, the full text of the IPPC Directive is reproduced in an annex.

Utrecht, March 1998

Table of Contents

Introduction

Ch. Backes[*]

The legislative process of adopting EC Directives more often than not involves a long and complicated history of negotiation and compromise. The result is that directives are frequently less clear and less consistent than most national legislation. Many questions remain and a lot of different interpretations are possible.

This certainly applies to the IPPC Directive.[1] A number of flaws in the IPPC regime must be recognized and must be solved by the authorities which have to transpose the Directive. Although most of the controversial issues have been settled during the drafting process, the result of this is not always sufficiently clear. Many questions discussed during the drafting process again arise as questions of interpretation in the current stage of transposing the text into national laws. The items and questions that were discussed at the 1995 Utrecht Conference have not yet been solved; moreover, they will become of direct practical relevance when the national laws that transpose the Directive have entered into force and are to be applied and interpreted in the light of the Directive by the national courts. In addition, these national courts, as well as the Commission, will have to consider the question whether these national laws meet all the requirements of the Directive.

The IPPC Directive was adopted by the Council on 24 September 1996, published on 10 October 1996, and entered into force on 30 October 1996. The Member States have until 30 October 1999 to take appropriate implementing measures. From around the publication date of this book, the discussion on what the Directive exactly requires and what the meaning of some of the provisions actually is, will become of practical relevance. The first evaluation report will have to cover the first three years (i.e. the period until 30 October 2002) of implementing 'national IPPC law', and is due to be submitted by April 2004. We cannot wait until then for an answer to at least some of

---

* Professor of Environmental Law, Utrecht University and Professor of European and International Environmental Law, University of Tilburg.

1 Council Directive 96/61/EC of 24 September 1996 concerning integrated pollution prevention and control, OJ 1996 L 257/26.

the many questions which remain concerning the Directive's various ambiguities and uncertainties.

The papers published here on the content and the context of the Directive, the approaches and interpretations in some of the Member States, and about an economic appraisal of the Directive aim to contribute to this discussion during and after the stage of transposition. In this Introduction the reader will find a brief overview of the main topics.

I The 'Best Available Techniques' (BAT)

According to Article 3 of the Directive, one of the general principles to be complied with by the operator concerns the best available techniques (BAT). Although this notion is defined in Article 2(11), many questions remain. Much of the discussion concerns the element 'available'. Explicitly, Article 2 envisages that 'costs and advantages' are to be taken into consideration; also, the techniques must be 'economically viable'. Macrory points out that both the components of economic viability and balancing of costs and environmental benefits which used to be included in the concept of 'best *practicable* means', has now been transferred to the notion of 'available' as a part of BAT. It follows that in substance no fundamental changes are introduced by the Directive in this context. Likewise, Emmott concludes that little difference can be observed between the requirement of BAT and the concept of BATNEEC (best available techniques not entailing excessive costs), which is the prevalent criterion in the UK and featured in earlier drafts of the Directive.

In Germany, however, according to Winter, a higher standard seems to apply, which fits in better with the precautionary principle than the IPPC definition does. These higher standards at least exist concerning certain effects of activities (*Gefahrenabwehr*). The author notes that this stricter approach can continue to be applied under the terms of Article 130T EC Treaty. However, the question arises whether *de facto* pressure will be exerted to level down these stricter requirements. It may be the case that, undesirably, such a move is even facilitated by Article 16 of the Directive which envisages publication of the measures taken in each Member State with respect to limit values relevant to certain categories of activities.

In fact, it would seem to follow from the general duty of interpreting directives in the light of the Treaty that BAT must be construed in conformity with the precautionary principle of Article 130R EC Treaty. The other Member States and the EU itself should thus follow the German approach rather than the other way around. Consequently, where there is a risk of significant damage to health and the environment, economic viability should not be taken into account. The permit should not be issued at all or under conditions which preclude this risk in order to comply with the Treaty's precautionary principle. By contrast economic viability may be taken into consideration in the setting of emission limit values where no risk of significant damage to health and the environment is present.

A further question regarding BAT is whether one standard applies throughout the whole Community or whether BAT may differ depending on local circumstances. According to Faure and Lefevere, regional cost-benefit comparisons should be decisive. Local advantages for the environment must be weighed against the local costs of pollution. Significant differences in the levels of BAT would then occur. Also Emmott takes this view, with reference to the *travaux préparatoires* of the Directive. Macrory, on the other hand, argues that know-how regarding various techniques may be exported so that eventually the same level of protection should apply Community-wide. The Directive itself is not clear on this point. But the wording of Article 16 seems to imply one BAT standard as the ultimate goal.

It is conceivable that a balance could be stuck as follows. From the perspective of each individual environmental compartment, the same level of protection should apply throughout the whole Community. Taking into account both the precautionary principle and the stand-still principle, it is not acceptable that in some areas of the Community more pollution, of for instance the atmosphere, is allowed than in others for the sole reason of the air being cleaner there. Local circumstances are, however, determinative where a balancing between the different types of pollution is concerned (air, water etc.) It follows from the very notion of integrated pollution control that a permit not simply reflects the sum of the different compartmental requirements but depends upon the overall effect of the activity in question on the whole of the environment. In one area of the Community, for instance, it may be less pressing to eradicate air pollution and to give priority to the reduction of water pollution as much as possible. In another part of the Community the reverse may be the case. On balance, the yield of protective measures in terms of environmental benefits with respect to water pollution may be less than with respect to air pollution. Even economic factors, such as the price of water, may be taken into account. Only with respect to such a balancing exercise should local circumstances be taken into account in deciding what constitutes the required best available technique.

A further unclarity concerning BAT is its relationship to prescribed emission limit values. Emission limit values must be seen as a minimum, according to Emmott. The BAT standard may never be allowed to overrule emission standards.

II BAT and BPEO

The standard of 'best available techniques', on the one hand, and the aspiration to give an integrated appraisal of all the environmental effects of the operation of an installation, on the other, raise the issue of the connection between BAT and the principle of the 'best practicable environmental option' (BPEO). This principle is particularly well-known in English environmental law. The Directive purports to adopt solutions which produce the best overall results in terms of environmental protection. As said, this totality involves more than a mere addition of the different sectoral BAT standards. Indeed, the fundamental new element of the IPPC Directive – compared to the previously compartmentalized EC environmental law – is the attempt to bring together the emissions to the air, water, soil, the production of waste, the use of energy and other environmental effects in order to achieve the best solution on balance.

In essence, however, a comparison between 'apples and pears' would often be required. For how does one decide upon a choice between a maximum reduction of either air or water pollution or the reduction of waste? The Directive itself does not give much guidance in addressing such a BPEO. Macrory reports on the experience in Britain since 1991. It became apparent that a purely numeric approach was not possible. In the final analysis, the search for the BPEO remains a process of weighing up certain factors. The main improvement of the IPPC Directive's 'holistisic' approach is that the required balancing between the various environmental effects must be carried out explicitly and more transparently. The English experience as noted by Macrory is helpful, although there is still no definite solution.

One lesson is that BPEO cannot be determined in general and in the abstract. Local and economic factors are relevant. Even where the same (type of) installations are involved, BPEO may differ depending on the area within the Community where they are established. It thus seems doubtful whether a full judicial review of the concept is possible. In practice the level of review of environmental permits within the Community varies considerably from one Member State to the next. The Dutch courts, for example, are not reluctant to review permits, allowing the administration little discretion, whereas the English courts allow a wide margin of appreciation to the authorities who originally issued the permits to determine what constitutes BPEO. In the light of the requirement of BPEO, however, the weight given in the case law of, amongst others, the Netherlands to (sectoral!) non-statutory technical standards (in e.g. circulars, guidelines etc.) is questionable.

III Coordination or Integration?

Next to BAT and the emission limit values one of the topics most discussed in the following papers is the integrated nature of the IPPC permits. In the contributions of Macrory, Drupsteen and Winter it is examined how the issuance of permits has evolved from a purely sectoral to a coordinated or even integrated approach, at least in the UK. Great Britain seems to be a bit further advanced in developing a real integrated system of permitting than the other countries. However, the protection of surface waters in the past was mostly kept to one side. Sweden was one of the first countries with a 'fully' integrated system.

At the EC level the integrated approach and the idea of an integrated permit appeared for the first time in 1987 in the 4th environmental programme, as Emmott points out. As for the IPPC Directive, does it prescribe a single permit or does it allow the coordinated issuance of several permits for one installation? Although Drupsteen expresses some doubts about whether the Directive would allow both approaches (as indeed other writers have done), Emmott and Winter conclude that the Directive does not prescribe a single integrated licence. Emmott supports this view with an analysis of the *travaux préparatoires*. 'Fully coordinated', the term used in the Directive, does not mean 'integrated'. Even the model of a 'supreme competent authority' did not advance beyond the drawing board. The Directive is now satisfied with coordination at a fairly superficial level (Macrory, Winter). This means a substantial backward step compared to previous drafts. Indeed, it is even doubtful whether the Directive actually covers all the environmental effects. This currently seems uncertain. According to Macrory, the visual impact of an installation, for instance, is not included.

It follows that it is very much an open question whether the finally adopted minimal solution of the Directive gives enough legal certainty so that the assessments will go beyond a mere assembly of generalised permit conditions. However, Winter argues that 'fully integrated' is not necessarily the best environmental solution. Cooperation between a number of authorities – each responsible for a different environmental compartment – has advantages over the situation where there is a single institution with competence for the whole environment.

IV The IPPC – Environmental Impact Assessment and Planning Law Relationship

How does the integrated or coordinated environmental licensing relate to the environmental impact assessment (EIA) and planning permission? An extra complication in this context is that the national legal systems are very different. Whereas EIA in some jurisdictions functions for the benefit of environmental permitting; in others, such as the UK, it is exclusively connected to the instruments of town and country planning. The problem with the latter system is that the data produced in this type of EIA (planning-related) do not meet the requirements of an EIA as prescribed within in the context of an IPPC permit. In particular, according to Macrory, they are insufficiently detailed. It would seem to follow that the IPPC Directive will have far-reaching consequences for English EIA law. More generally speaking, little consideration has been given to the coordination of the IPPC process with permits in the context of planning law, despite the obvious practical relevance (Macrory).

V The Relationship between the IPPC Directive, other EC Environmental Law and other Kinds of Instruments

In his keynote-address, Peters stresses that the environmental law of the EC is compartmentalized, badly structured and complicated. There remains much to do along the way to a simplified and more coherent system. The IPPC Directive is an important step on this road, but it is only the first step. The adoption of the Directive has illustrated that it forms merely one ingredient of a broader, multifaceted integrated policy, both from a Dutch and from an EU perspective.

In several of the papers, the remarkable feature is noted that the IPPC Directive is rather outmoded: it exclusively reflects a command and control type of legislation. An approach which sits uncomfortably with a general trend towards more self-control and self-responsibility, prevalent for some years at both national and EC level (Winter, Jans). The fifth environmental action programme, for example, envisages a mix of instruments, including – but amongst several others – the prohibition and permit system. The IPPC Directive practically ignores this development. Macrory highlights the fact that it is merely in the context of an integrated best practicable environmental option that command and control have not worked. At present, a more successful approach involves consultations with the (future) operator. In the Netherlands, moreover, voluntary agreements or covenants between the authorities and operators or their organisations have become an important instrument.

One important issue for consideration by the Member States' legislatures during the transposition of the Directive into national law is whether and to what extent the Directive leaves some scope for the application of other instruments and for less detailed, more open permits. On the one hand, as noted by Drupsteen, Article 9(3) stipulates that the permit must contain emission limit values for certain pollutants for the protection of the air, water and land. On the other hand, Article 9(4) forbids the prescription of specific technologies or techniques. Therefore, one could reverse the question, stressed by Drupsteen, as to whether the Directive precludes the development of more flexible, global permits, as currently adopted in the Netherlands: does the Directive, in particular Article 9(4), in certain circumstances, even require less detailed permits? After all, specified techniques cannot be made mandatory. At least in a number of Member States this is frequently the case.

Another question in this context is whether the Directive entails only duties for the authorities or for operators as well. That is, obligations directly binding upon private parties without any need for them to be made explicit in a permit. According to Macrory this is indeed the case. According to Emmott, however, the Directive lays down no unequivocal hard and fast rules and contains contradictory texts at the level of general principles. An operator does not breach the law if he complies with his permit, even if he acts contrary to the general principles laid down in Article 3 of the Directive. He is not directly bound by them. Jans underlines the fact that this result is a backward step compared to the case law of the Court of Justice, causing difficulties with the enforcement of the Directive. Winter, on the other hand, thinks that Article 3 includes an extra unwritten duty. This is a first step, uncommon to EC law but familiar to German law. Other such duties, and more explicit ones, may follow. Interpretation of this provision of the Directive after transposition into national environmental law promises to be interesting.

Another also interesting question is of how dynamic the national permit system based on the IPPC Directive is allowed or has to be. The Directive is not clear in that respect. How often and under which circumstances are the licences to be updated and what rights can the operators derive from the permits?

All these questions show that the discussion on the interpretation of the IPPC Directive and its consequences for the national legal systems was far from complete when this important piece of EC environmental legislation was published; indeed to a great extent the discussion has just begun. We hope that this book will function as a small contribution to that discussion.

The IPPC Directive in the EC Environmental Law Context

Environmental Framework Directives of the European Union – Ideas on Coherence and Suggestions for a Logical Basic Structure

*J.A. Peters**

I Introduction and Summary

The coherence between the EC framework Directives on environmental management needs to be improved. At the same time, a logical basic structure should be established. It is essential that the present patchwork of Directives be transformed into a simpler and more transparent system, in order to enhance the practicability of the European legislation.

This would boost European efforts to achieve a high degree of environmental protection and sustainable development, without changing the protection targets as such. It would also constitute a step towards the creation of a level playing field for companies in the different Member States.

The elements presented below are intended to open the debate. Our most important task now is to discuss the outlines of the system; details can be filled in at a later stage.

It is by no means our intention to disrupt the current legislative process. The Commission has put forward a number of proposals to which we adhere where possible. Consultations on the IPPC Directive are already at an advanced stage, a framework Directive on air quality and a Directive on ecological water quality are being debated and an action programme for groundwater is being drawn up. This is a good opportunity to exchange ideas on the interrelationships between these proposals and the basic structure underlying these Directives.

The proposals below would lead to a number of alterations in the proposals put forward by the Commission. A number of existing Directives might in due course be

---

* Director for International Environmental Affairs, Ministry of Housing, Spatial Planning and the Environment.

incorporated into framework Directives, or be transformed into 'daughter' directives. We should like to point out, however, that in principle the ideas are compatible with our earlier positions on the Directives in question.
The proposals basically entail translating a number of principles – on which there appears to be a large degree of consensus – into concrete actions:

- Deregulation. The proposals would mean a reduction in the number of Directives and the creation of a transparent structure, resulting in the elimination of overlapping and unnecessary provisions.
- Subsidiarity. Structuring the system of Directives would at least create clarity regarding the dividing-line between the Community's responsibilities and those of individual Member States. It is likely that the number of specific subjects regulated by the Community would decrease on balance, particularly in the case of instruments for achieving targets and norms.
- Coherence. The sectoral legislation would be integrated, in line with the aims of the EU's fifth Environmental Action Programme.

Since the first Environmental Action Programme was published in 1973, dozens of EC environmental Directives have been adopted. They have exerted an ever-increasing – and beneficial – influence on the form and content of the environmental policies of the individual Member States. However, twenty years of experience in implementing the Directives have brought many problems to light – problems in the very areas mentioned above. Legislation is sometimes too complex or detailed, sometimes takes too little account of the extent to which Member States are able to tackle their problems in their own way, and often display a lack of cohesiveness. It is important that these problems are addressed now, while the 'second generation' of environmental Directives is being drafted.

A few attempts have been made to impart some coherence (e.g. Council Directive 91/692/EEC on standardizing and rationalizing reports on the implementation of certain Directives relating to the environment and more recently the proposal for the IPPC Directive). In general, however, the coordination of policy on air, water and groundwater/soil, needs improvement.

The backbone of the proposed new structure would consist of four central Directives, that interrelate and support each other, namely a Directive on the integrated issuing of permits to large companies, and a framework Directive on each of the three environmental compartments (air, water and groundwater/soil). The IPPC Directive on integrated licensing (in combination with general rules) for large companies could concentrate on emission standards based on Best Available Technology (BAT), on emission limit values at EU level for major processes and substances (if necessary) and on lists of substances. The framework Directives for air, water and groundwater/soil should each be tailored to suit the specific characteristics of the compartment in question. They might focus, for instance, on the quality of the compartment, on programmes of measures for all emissions (large, small, diffuse), on Best Environmental Practices (BEP) and on monitoring programmes.

Council Regulation 793/93 on the evaluation and control of the risks of existing substances could be used as an additional, fifth pillar of this system, in order to foster integration from the point of view of environmental assessment of priority substances.

This proposal does not encompass the subject of nature conservation, which occupies a separate position in the EU environmental legislation. The scope for streamlining EU nature conservation measures and their relationship with the framework directives discussed in this proposal require further examination.

The proposed structure would resolve a number of important issues:

- the relationship between emission targets and quality objectives;
- the set of instruments available for reducing emissions (including licensing and general rules);
- lists of substances and risk assessment;
- diffuse pollution and the links between compartments.

This is also a good opportunity to critically review the role of the Community and that of the individual Member States in the making and implementation of environmental policy. This aspect will become more important as the EU expands. The Member States should pursue independent policies when it comes to problems within their own borders, and groups of Member States can be encouraged to work together to tackle transboundary problems that can be effectively addressed at regional level (such as in river basins). Community action is necessary with regard to transboundary forms of environmental pollution, or to ensure the harmonisation of techniques for combating the types of pollution that occur in most Member States.

Agreements and rules on long-term environmental quality objectives (representing a level that safeguards sustainable development) and short-term objectives (a minimum level of protection) can be set at national, regional or community level, as appropriate.

II Proposed Basic Structure

In order to make clear what changes will be necessary, we shall first describe a possible 'ideal' basic structure. Section C then takes a closer look at a number of issues that touch upon all the Directives, and discusses some of the gaps in the outline presented below, particularly the question of whether emissions not covered by the IPPC Directive should be limited with the use of Community instruments. Although it is important that policy with respect to the different compartments (water, air and groundwater/soil) should be drafted along similar lines, some differentiation might be appropriate. The dispersal of substances in the atmosphere, for instance, is different from that in water, which in turn is different from the dispersal that occurs in soil and groundwater. The traditional policies for the three compartments also differ, and the policy instruments chosen must take account of these differences.

1 *Framework Directive on Water*[1]

(To be drawn up on the basis of the proposed Directive on ecological water quality, using elements from Directive 76/464/EEC and existing water quality Directives).

Main instruments:

- quality objectives
- programmes of measures for all discharges (large, small, diffuse) into river basins and regarding hydrology, morphology and biotic communities
- lists of substances
- BEP (Best Environmental Practices)
- monitoring programmes per river basin

Quality objectives will be set as follows: (1) at the level of individual Member States specific to the local situation ('operational' objectives for river basins or drainage areas); (2) at EU level for specific functions of surface waters, binding in so far as Member States attribute the function in question to waters (cf. existing EC Directives); and (3) at EU level (in due course, if it appears necessary and on the basis of experience gained at national or regional level), in the form of generally applicable minimum quality levels, as a recommendation rather than a binding rule. The quality objectives referred to under (1) may also be set by two or more countries jointly (cf. the '*Zielvorgaben*' for the Rhine).

The first set of objectives, which may be geared towards safeguarding specific functions and/or nature protection, refer to the chemical quality as well as the ecological integrity of surface waters. They concern, therefore, not only substances but also the hydrological, morphological and biotic conditions. As far as hydrology is concerned, the objectives for groundwater and soil will also be taken into account.

This Directive must also include a provision for the indirect protection of soil and groundwater by, for instance, incorporating standards for suspended material and/or aquatic sediment in the quality objectives.

A programme of measures for a river basin is to be drawn up by the Member State or States concerned, and specifies the means by which the quality objectives are to be achieved. These may well include measures other than legally binding instruments. As regards substances, the programme will specify the necessary reduction of all emissions into water (including diffuse pollution) which must be achieved within a certain period of time. The programme takes account of the obligations geared towards controlling emissions (mandatory licensing, BAT) to which large companies are subject under the IPPC Directive (see § 4).

1 The term 'Framework Directive' is used in this section to indicate the central role of the Directive in question within the system of environment Directives. It is no indication of the presence or absence of material provisions in the Directive.

2 *Framework Directive on Air*

(To be drawn up on the basis of the proposed Directive on ambient air quality assessment and management).

Main instruments:
- quality objectives
- critical loads
- programmes of measures for all emissions (large, small, diffuse) per area
- lists of substances
- BEP (Best Environmental Practices)
- monitoring programmes per area

A generally applicable minimum quality will be set at EU level to serve as a binding quality objective. The 'critical loads' instrument may also be applied. The loads are based on critical values for the soil and, in some cases, for surface water (see also § 3). It will probably prove possible for a number of Member States to agree on a critical load for a large area (cf. the ECE protocol on SO_2, Oslo 1994). Critical loads link air pollution to the subsequent pollution of soil and water and are meant to safeguard the quality of the latter two in so far as it is affected via the air.

A programme of measures for an area is drawn up by the Member State or States concerned and specifies the means by which the quality objectives or critical loads are to be achieved. These may well include measures other than legally binding instruments. The programme should also indicate what reduction in overall emissions to air is to be achieved within a certain timeframe.

The programme takes account of the obligations geared towards controlling emissions (mandatory licensing, BAT) to which large companies are subject under the IPPC Directive (see § 4).

3 *Framework Directive on Groundwater/Soil*

(As yet there is no Commission proposal that can be used as a basis; the groundwater Directive – 80/68/EEC – and the forthcoming groundwater action programme may provide elements for a new Directive).

Main instruments:
- critical loads
- programmes of measures per area
- lists of substances
- BEP (Best Environmental Practices)
- monitoring programmes per area

Unlike the procedure in relation to water and air, no quality objectives will be set for soil. Instead, critical loads are to be defined.

A critical load is the maximum acceptable amount of a certain substance that can enter the soil per unit area and per unit time, either directly (e.g. from fertilisers) or indirectly, via the air (e.g. from acidifying substances). The critical load is derived from an acceptable concentration (critical value) in the soil, using a model. The critical value is similar to a quality objective, but does not have the same status. This is because, in contrast to water and air, soil that is heavily polluted does not in the short term become cleaner as a result of dilution or decomposition.

Critical loads can be established for a large area (several Member States or even the entire EU) in the case of airborne deposits (e.g. acidification); see also § 2. However, frequently it will be the direct local load that is crucial, in which case it would obviously be best for the Member States themselves to set the critical loads. The acceptable concentrations mentioned earlier should preferably be set at Community level (as a reference framework rather than a binding standard).

Critical loads would have to take explicit account of the standards for surface waters, for instance with the help of calculations on the amount of leaching to deeper groundwater and surface waters.

A programme of measures for an area is to be drawn up by the Member State or States concerned and specifies the means by which the critical loads are to be achieved. It will include measures to reduce the pollution of the soil from different types of land use (agriculture, industry, transport, housing etc.) and with regard to the storage of waste on the soil surface.

Not only the local deposition of substances (e.g. from the application of fertilisers and pesticides, leakage from factories) must be taken into account, but also diffuse pollution that results from, for example, tyre wear and tear and oil loss from road traffic, and the corrosion of road signs, guttering and other metal structures. The programme should take account of obligations geared towards controlling emissions (mandatory licensing, BAT) to which large companies are subject under the IPPC Directive (see § 4).

As with surface waters, targets should be set not only for substances but also for the hydrological, morphological and other geological conditions (to protect from water depletion, for instance) to safeguard the ecological integrity of the soil. Such targets can best be determined locally and must be realised partly by using instruments related to the quantitative management of surface waters.

4 IPPC

Main instruments:

- integrated licensing (in combination with general rules) for large companies
- emission standards based on BAT in permits (or general rules)
- exchange/harmonisation of BAT and emission standards and, if necessary, emission limits at EU level for major processes/substances
- lists of substances

Under this Directive emission standards based on BAT will be incorporated into permits issued to large companies. A certain degree of harmonisation will occur as a result of the exchange of information at EU level as regards the techniques and emission standards prescribed by individual Member States.

A Community minimum level for emission standards might prove desirable in the case of transboundary pollution and/or excessive discrepancies between Member States. If so, emission limits aimed at a high degree of protection would be laid down in EC Directives.

It is possible that permit conditions based on limit values applicable at Community, and perhaps at national, level, would in certain situations not be strict enough to achieve the quality objectives for water, soil or air in a particular area. (The quality objectives can be those contained in existing EC Directives, such as the fresh waters for fish Directive or the NO_2 Directive, or they could have been laid down as part of a programme as referred to in §§ 1–3.) *In this case additional measures would need to be imposed on the companies in question.*

The emission standards laid down in the framework of the IPPC Directive would therefore serve as a *basis* for the programmes to be drawn up, as far as industry is concerned, but not as an upper limit. It will remain important for Member States to examine whether these emission standards make a sufficient contribution to the realisation of the quality objectives, or whether additional requirements must be introduced for 'IPPC companies'.

5 Council Regulation on Existing Substances (793/93)

(In contrast to the Directives referred to above, this is an *existing EU regulation not undergoing amendment.* The description below sets out what elements in the existing regulation are most important for the proposed basic structure and what action could be undertaken on the basis of the regulation to enhance integration between the environmental directives).

Main instruments:

- selection of priority substances
- uniform risk assessment of substances
- provision of information by companies
- keeping the records on existing substances up to date
- proposals for risk control

The list of priority substances drawn up annually on the basis of this regulation (the first series contained over 40 substances) should be fed from the lists of the framework Directives referred to above. This would result in a harmonisation of the lists, without all lists necessarily being identical.

This regulation should also be used to support the setting or amending of quality standards for water, soil and air, in applying one single method of risk assessment. This would produce maximum permissible concentrations of each substance in water, air and soil, which could be used as a basis for determining quality objectives or (in the case of soil) critical loads.

III Important Issues Related to the Proposed Structure

1 Relationship Between Quality Objectives and Emission Standards

Experience with the implementation of existing EC Directives (such as 76/464 and the water quality Directives) and of other international agreements (such as the Rhine and North Sea Action Programmes) has shown that it is extremely difficult to establish a correct numerical relationship between (1) a quality objective for a substance and (2) emission limits or permit conditions for the same substance. This is in part due to the fact that the quality of a particular body of surface water or volume of air or soil is usually affected by several sources of pollution at the same time, for which emission standards are not all set simultaneously. This can be seen most clearly in the case of direct discharges as opposed to indirect or diffuse discharges.

The only international framework which has been relatively successful in numerically linking emission targets and quality objectives is the Rhine Action Programme (RAP). The success of the RAP approach can be traced to two factors: (1) the programme covers a single, cohesive river basin, and (2) both direct and diffuse sources of pollution were examined at the same time. (The RAP procedures were not, it should be noted, based on formal legal requirements).

A similar method would not be appropriate for the EU as a whole, since the different geographical regions of the EU are not sufficiently cohesive. But it could be used for individual areas (such as a river basin), in line with the Commission proposal for the directive on the ecological quality of water. This would mean that a numerical relationship between quality objectives and emission standards (including emission reduction objectives) could be defined *only* in those programmes that have to be drawn up per area. This applies not only to surface waters but also to soil and air. (The definition of the areas need not be the same for all three compartments). Consequently, *no* numerical relationship can be defined between quality objectives and emission requirements at EU level.

In terms of emission policy (the 'source-based approach'), this means that the IPPC Directive should only include provisions on BAT and emission standards. As far as the relationship between these emission standards and quality objectives in the framework Directives on water, air and soil are concerned, it would be enough simply to include a provision in the IPPC Directive stating that stricter conditions must be incorporated into a permit if the relevant quality objectives cannot be achieved using BAT. In terms of environmental quality policy (the 'effect-based approach'), this means that the framework Directives on water and air would not define a direct relationship between a minimum quality to be established at Community level (which is desirable for air; in relation to water this still has to be considered) and the above-mentioned emission standards. The Community emission standards should be applied even if the minimum quality or the quality objectives for the individual area have been achieved.

At Community level, the source-based and effect-based approaches would exist side by side, with no (numerical) relationship between them. However, for individual areas attempts should be made to define a numerical relationship between the applicable quality objectives and emission standards. This relationship could be defined either by an individual Member State, or by several Member States jointly, as appropriate. Naturally, they would have to be at least as strict as any minimum emission standards or minimum quality set at Community level.

2 *Instruments for Reducing Emissions*

The IPPC Directive introduces a system of permits and general rules for large companies. One issue connected with this Directive is the need for harmonisation between the licensing policy (with emission standards) in the Member States and the possible introduction of Community emission limits. It is important that the licensing measures of the Member States can be introduced immediately, without the need to wait for such Community standards.

No harmonisation of instruments for reducing emissions is envisaged for small companies/installations. It is not possible at this stage to answer the question of whether it would be worthwhile harmonising instruments at EU level, but the matter must be discussed. (At the moment, for example, there is a fairly broad licensing obligation for discharges to water based on Directive 76/464: should this be continued, and if so, should it be harmonised with the rules for air and groundwater/soil?) The discussion should be guided by the subsidiarity principle and the need for deregulation. It should cover all possible instruments, including fiscal instruments and voluntary agreements as well as permits and general rules, and the question of whether any harmonisation should apply to all categories of polluters or only some should also be addressed.

If the decision is taken to harmonise, various systems of regulation would be possible. One possibility would be to include in the framework Directives on water, air and groundwater/soil provisions regarding emission reducing instruments. The advantage of this would be that the regulations could be specially tailored to the different compartments. Another possibility would be to introduce one separate directive harmonising the rules for small companies/installations in relation to water, air and groundwater/soil. The advantage of this option would be a simplification of the system.

3 *Lists of Substances and Risk Assessment*

All the Directives discussed here are based on lists of substances. It is highly important that the substance lists for water, soil and air be coordinated as far as possible. In principle this could be done within the framework of the IPPC Directive. The drawback of this, however, would be that some substances would be overlooked since they do not figure in the emissions produced by large companies.

It would be better to draw up the lists at compartment level, since it is here that the problems are most visible. They could be updated on the basis of the records kept under the prioritisation system of the Regulation on existing substances (793/93).

Conversely, the lists for water, soil and air can be taken into account when the list for the Regulation is drawn up.

We would recommend that one system of risk assessment be applied to all three compartments, both for the Community minimum quality, and for regional quality objectives. The method could be based on the system devised in the context of the Regulation on existing substances (793/93), although specific methods would have to be introduced for certain substances (e.g. oxygen and nutrients in surface waters). This would not mean, incidentally, that the standards themselves would have to be the same at all levels. After all, standards are based not only on risk assessment but also on policy choices regarding deadlines, economic consequences etc.

In other words: the scientific method should be harmonised, but the standards may be differentiated as a result of policy-related considerations.

For the record, we should also refer to Directive 67/548/EEC, which would come into play as soon as *new* substances came into the picture.

4 Diffuse Pollution and Links Between Compartments

The scale of diffuse pollution relative to direct emissions is increasing, especially in the case of surface waters. It is therefore important that the new structure take better account of this type of pollution than is the case in the existing Directives. However, the problem of diffuse pollution often makes itself felt first in another compartment, such as soil pollution borne by the air, or the pollution of surface waters via the soil.

In the new structure, diffuse pollution must therefore be tackled by measures at source i.e. provisions should be made in the Directive that refers to the compartment that is first affected. We must therefore ensure that a framework Directive for compartment 'X' also guarantees protection for compartments 'Y' and 'Z'. The outlines for the framework Directives on water, air and soil given in section B include a brief description of the necessary provisions. The details require further elaboration. It is also important that legislation relating to substances and products takes as much account as possible of the distribution of those substances through different compartments. Examples include the Regulation on existing substances and the Pesticides Directive (91/414/EEC).

The relationship between the compartments is also affected by factors other than pollutants. Water flows and water levels in the soil, underground aquifers and surface waters are closely linked. The Netherlands is therefore of the opinion that the framework Directives on both water and groundwater/soil should take sufficient account of this interaction.

IV Implications for a Number of Existing or Forthcoming Directives

The basic structure outlined above will have implications for the content of a number of existing and proposed Directives. The most obvious changes are summarised below.

No suggestions are made as yet with regard to the harmonisation of emission reducing instruments for discharges not covered by the IPPC Directive (see section C, § 2).

Apart from the Directives and Regulations mentioned below, there are many others that in some way impinge upon the proposed structure. Examples include the Directives on nitrates and urban wastewater, the quality Directives for surface water, the sewage sludge Directive and the wastes Directives. An inventory of these Directives and their impact on the structure would be useful.

1 *Directive on Ecological Water Quality*

- This would evolve into a framework Directive on water.
- The ecological aspect would be more strongly emphasised, i.e. clear provisions concerning the hydrology and morphology of aquatic ecosystems would have to be incorporated into the Directive, alongside the provisions on chemical quality.
- In due course, a minimum quality for surface water could be incorporated into the Directive. A decision on this could be taken after some initial experience has been gained with quality objectives applied at national or regional level.
- The current standards of the water quality Directives on 'surface water for drinking water preparation', 'bathing water', 'fish waters' and 'shellfish waters' could be retained (or tightened up), perhaps as 'daughter' Directives of the new framework Directive on water.
- A list of substances would be incorporated into the Directive, similar to the lists in 76/464; however, the difference between black and grey lists should be dropped.
- Elements of 76/464/EEC would be incorporated into the Directive (see § 2).

2 *Directive 76/464/EEC*

- The elements of this Directive relating to large companies (licensing, BAT, emission limits in 'daughter' Directives) would be transferred to the IPPC Directive.
- All other elements would be transferred to the Directive on ecological water quality. These are as follows:
  - measures for list II substances (licensing, programmes),
  - elements relating to quality objectives,
  - licensing of list I substances not covered by the IPPC Directive, and the existing 'daughter' Directives containing emission limits, in so far as they apply to companies not covered by the IPPC.
- This Directive could therefore eventually be withdrawn.

3 *Air Quality Directive*

- The quality standards need to be clearly defined, and must take into account the other environmental compartments (see also the critical loads for soil). Both Community and national standards can be set; the former would be binding.

- A start has been made on using models to define the relationship between air quality and emissions. Models should be refined and included in national programmes (measures are currently being discussed, but not programmes).

4 *Groundwater Directive*

- This would evolve into a framework directive on groundwater and soil.
- The Groundwater Action Programme could provide a basis for a new set-up.
- The Directive is still mainly emission-oriented. The quality-oriented element and hydrological aspects would have to be incorporated.
- The rules on 'indirect' pollution require clarification. Pollution caused by activities such as agriculture would also fall under the Directive. The system set up under the nitrate Directive could perhaps be used here.
- The strict distinction between 'groundwater' and 'soil' should be softened.
- The provisions in the Directive should take account of leaching both to deep groundwater and to surface water.

5 *IPPC*

- In view of the debate on IPPC, which has been going on for over a year now, this Directive is under the greatest pressure in terms of time.
- The relationship between emission standards and quality objectives must at any rate be properly regulated, to ensure that the Directive can be effectively incorporated into the basic structure. It should be specifically stated that stricter emission standards must be included in permits if quality objectives require it.
- There must be no introduction of a system that allows countries to replace Community or national emission conditions with a 'quality objectives approach'. In contrast to the present situation under 76/464, both instruments would have to be applied, with a numerical link between emission conditions and quality objectives set at regional level, rather than at Community level (see section C, § 1).

An Overview of the IPPC Directive and its Development

N. Emmott[*]

I Introduction

Directive 96/61/EC on IPPC was adopted on 24 September 1996, three years and ten days after it was first formally proposed by the Commission. By the usual standards of EC environmental law, this is not a long time for negotiation of a new Directive. Indeed, given the importance of the Directive in marking a shift from single-medium to multi-media legislation, an interval of just over three years between the tabling of a formal proposal and agreement on a final legal instrument appears quite short. What these dates hide, however, is the fact that work towards the establishment of what ultimately became the IPPC Directive started some seven years before the Commission's formal proposal was brought forward. Therefore, the 'lead time' between the idea of the Directive and the reality of its adoption should be seen as approximately 10 years. The concepts and practices which underpinned the original suggestion of the Directive evolved over a longer period still.

This chapter seeks to provide an overview of the IPPC Directive, and of the processes and influences which led to its adoption. The overview of the Directive itself will be of interest to those readers who simply wish to know the main requirements and provisions which arise under the EC system of IPPC. This information is presented primarily in the form of a factual summary, but also includes some analysis and commentary on the requirements and structure of the IPPC regime. The discussion of the processes and influences which led to the Directive, meanwhile, is intended to offer some insight on the context in which the EC system of IPPC has been created, and thus to illustrate that the Directive forms just one ingredient of a broader, multi-part philosophy of integrated environmental policy.

* Research Fellow, Institute for European Environmental Policy, London.

The expression 'IPPC' – Integrated Pollution Prevention and Control – is sometimes used interchangeably with the term 'IPC' – Integrated Pollution Control. The two phrases can be considered to have the same meaning if the control of pollution is taken to include prevention.[1] For consistency, the term IPPC is used in the text except where discussing systems or initiatives which have explicitly applied the heading 'IPC'.

II The Development of IPPC in a Broad Sense and a Narrow Sense

At the outset of this discussion, it is important to note that the expression 'Integrated Pollution Prevention and Control' has a broad sense and a narrow sense, although the two are inter-related. The IPPC Directive is an example of the narrow sense, being concerned with the integrated control of polluting emissions and other environmental impacts from specific industrial activities. In contrast, the broad sense of IPPC reflects a general philosophy based on the premise that pollution problems should be addressed taking account of all three environmental media – air, land and water – in an integrated manner. This contrasts with the more traditional approach towards pollution control of focusing on one medium at a time, which can have several disadvantages: it may result in the transfer of pollution to the other media; it may overlook the biogeochemical and commercial cycles of pollutants; or it may fail to consider multiple exposure paths for humans and the environment.[2] The broad IPPC concept aims to overcome these problems and can be put into practice through controls applied to polluting sources, polluting substances, or geographical areas which are experiencing pollution.[3] The IPPC Directive is thus just one of many possible forms through which the broad sense of IPPC may be given expression.

The application of the IPPC concept to sources of pollution can relate to various areas of control. Sources can include, for example, industrial processes or installations, products, or economic sectors as a whole such as energy, agriculture or transport. When used to regulate industrial processes or installations – which is the purpose of the IPPC Directive – the adoption of an integrated approach can help to support the introduction of clean technologies rather than 'end-of-pipe' solutions. This is because an end-of-pipe approach, such as filtration or scrubbing of polluted gases, may produce a solid or aqueous waste. The IPPC approach, on the other hand, should consider opportunities for substitution of raw materials or process changes, in order to prevent the formation of the polluted gas in the first place.[4]

---

1 N. Haigh and D. Wilkinson, *Integrated Pollution Control* (1993), WWF International Discussion Paper, p. 3.

2 OECD, *The State of the Environment*, Paris (1985), p. 264.

3 F. Irwin, 'Introduction to Integrated Pollution Control', Chapter 1 in: N. Haigh and F. Irwin, eds., *Integrated Pollution Control in Europe and North America* (1990), The Conservation Foundation, Washington DC and Institute for European Environmental Policy, London, p. 11.

4 Ibid.

The suggestion that industrial sources of pollution should be subject to some type of integrated control, at least at an institutional level if not a functional one, is by no means a modern idea. Over one hundred years ago, in 1880, the first Chief Inspector of the Air Pollution Inspectorate in the UK asked the government 'to consolidate the departments of air and water so far as the chemistry is concerned and to report them together'.[5] It was many years, however, before any concrete steps in the direction of truly integrated industrial control would be taken. Sweden, for example, was one of the first countries to take significant action in this area when a single permitting process to regulate releases to air, water and land from large installations began to be introduced under the Environmental Protection Act of 1969.[6] Other countries have followed with integrating initiatives of their own, although in some cases these have involved integration of procedures or authorities without necessarily integrating the actual controls applied. Such steps may bring benefits of improved regulatory efficiency but will not necessarily overcome the inherent disadvantages of single-medium pollution control discussed earlier.

The region-based approach to IPPC aims to achieve a desired level of environmental quality in a defined area by controlling inputs via all media. By way of illustration, initiatives to improve water quality in the Great Lakes initially focused only on direct discharges to the aqueous environment, but have now recognised the need to look at deposition from the air and seepage through the ground as well. This followed from the determination in 1981 that some 90 per cent of PCBs in Lake Superior resulted from air deposition, and the subsequent finding that hazardous waste sites were additional significant sources of pollution. The Canada/US Great Lakes Water Quality Agreement, which sees the Great Lakes basin as an ecosystem comprising 'the interacting components of air, land, water, and living organisms, including humans', now calls for lake management plans to reduce the loading of critical pollutants from various sources. In Europe, a similar, region-based approach for pollution prevention and control is being applied by the nations whose discharges contaminate the North Sea.[7]

The substance emerged as another obvious focus for efforts aimed at integration because it is the harmful effects of substances or pollutants on receptors or targets which we wish to prevent. Substance-based approaches may be appropriate in respect of particularly hazardous, persistent or otherwise problematic pollutants. Ideally, regulation should take account of the possible harmful effects of a chemical, and different possible routes from release to exposure, before identifying control strategies.[8] From the 1960s onwards, national governments have introduced a significant range of controls on the manufacture, use, release, storage, transport and disposal of hazardous chemicals. Much work has also been undertaken at the international level, for example under the auspices of the OECD. In particular, in 1985 the OECD initiated a study

---

5 *Supra* note 1, p. 7.

6 D. Hinrichson, 'Integrated Permitting and Inspection in Sweden', Chapter 7, *supra* note 3.

7 *Supra* note 3, pp. 21–22.

8 *Supra* note 3, p. 15.

on the management of chemicals under the working title of 'Integrated Pollution Control'. This study was to prove an important catalyst in the events which finally led to the adoption of the IPPC Directive, as outlined below.

III 1986–1991: Development of the Idea of an EC Approach to 'Integrated Permitting'

The OECD chemicals working group undertaking the 'Integrated Pollution Control' project in 1985–86 involved participants from a number of organisations. Among the contributors were representatives from the Conservation Foundation in the US and the Institute for European Environmental Policy (IEEP). The Conservation Foundation had previously carried out a significant amount of work on 'cross-media pollutants',[9] while the IEEP had been active in the field of EC environmental policy. As the OECD chemicals study drew to a close, the two organisations established a cooperative project for further study of the topic of IPC (as it was then called), with financial assistance from the German Marshall Fund. This looked beyond the OECD's initial focus on the control of chemicals into other areas including the source- and region-based approaches, examining practices and experiences in several countries and at the international level.

The results of the Conservation Foundation/IEEP project were presented at a symposium held in Brussels in November 1988.[10] It was at this symposium, attended by officials from Directorate-General (DG) XI of the Commission that, among other conclusions, the idea of an 'integrated permitting' Directive was first proposed. A few months previously, IEEP had been commissioned by DG XI to conduct further work on the possible development of EC policy in this area. This led to the production of a report delivered to DG XI in 1989 which recommended, as a first priority, a Directive on integrated permitting.[11] This conclusion was based not only on the desirability of the integrated approach over single-medium routes, but also on the observation that several Member States were already moving in this direction. If EC policy did not follow the same path, it was observed, there was a danger that those countries which wished to take steps towards greater integration could be hindered by single-medium requirements specified at the EC-level.

---

9 Reported, for example, in publications such as *Controlling Cross-Media Pollutants* (1984) and *New Perspectives on Pollution Control* (1984), The Conservation Foundation, Washington DC.

10 The results of the study were also described in N. Haigh and F. Irwin, eds., *Integrated Pollution Control in Europe and North America* (1990), The Conservation Foundation, Washington DC, and the Institute for European Environmental Policy, London, which is widely referred to in this chapter.

11 N. Haigh, *Possibilities for the Development of a Community Strategy on Integrated (Multi-Media) Pollution Control* (1989), Report submitted by the Institute for European Environmental Policy to DG XI, October 1989.

The report's conclusions concerning the need for a Directive were accepted by DG XI despite resistance from some quarters. No doubt the response from DG XI was shaped by the fact that the general principle of the integrated approach had already been recognised in 'official' EC environmental policy through the adoption in 1987 of the Fourth Environmental Action Programme.[12] The Programme had noted that the tightening of environmental standards for one medium leads to increased pressure on the other media, and while it was not definitive in its support for the integrated approach, it stated that it was 'at least arguable' that a single-medium approach was not necessarily the most efficient.

Having accepted the principle that a Directive should be proposed, DG XI subsequently placed activity on hold in order to give priority to the proposal which ultimately led to Regulation 1836/93[13] on a Community eco-management and audit scheme. Drafting of the proposal which was to become the IPPC Directive therefore did not get started until 1991. In the interim, however, significant events were proceeding elsewhere, notably in the UK and under the auspices of the OECD.

In 1989, building on its previous work in the field of chemicals control and other activities, the OECD started a broader study on the subject of integration. This was to lead in 1991 to the adoption of a Council Recommendation on IPPC,[14] which remains one of the widest and most complete official expositions of the broad sense of IPPC produced to date. The Recommendation advises OECD member countries to:

> practice integrated pollution prevention and control, taking into account the effects of activities and substances on the environment as a whole and the whole commercial and environmental life cycles of substances when assessing the risks they pose and when developing and implementing controls to limit their release.[15]

An Appendix to the Recommendation sets out guidance concerning, *inter alia*, foci for decision making, legislation, and management instruments. The discussion of foci for decision making identifies the possibilities of concentrating on the substance, the source and the geographical region. Source-based approaches are further delineated as potentially relating to industrial processes, products and economic sectors.[16] As regards legislation, the recommendation is that integration be pursued not only through specific environmental laws, but also through laws in related areas such as energy, transport and taxation.[17] It is additionally suggested that integration can be achieved

12 EEC Fourth Environmental Action Programme (1987–1992), OJ C 328, 7 December 1987, paragraph 3.2.3.

13 OJ L 168, 10 July 1993.

14 Recommendation of the Council on Integrated Pollution Prevention and Control, C(90)164(final), adopted 31 January 1991.

15 Art. I(a).

16 Appendix, paragraph 3.

17 Appendix, paragraph 4.

through a variety of management instruments, one of which is single permits for polluting sources.[18]

In parallel with the OECD work, steps were being taken in the UK to introduce a system for the integrated permitting of the most complex industrial processes. The framework for a regime, which was introduced under the heading IPC, was put in place by the Environmental Protection Act 1990, and the system was phased-in for specific industrial sectors in the years which followed. This UK initiative was in part a response to comments made by the Royal Commission on Environmental Pollution as long ago as 1976 on the need for a more integrated approach to industrial pollution control.[19]

By early 1991, therefore, the idea of an EC Directive had been accepted in principle but remained on hold in practice, the UK (among others) had introduced a regime which in many respects resembled the anticipated Directive, and the OECD had recommended application of the broad IPPC concept including for the authorisation of industrial plants. Work at the EC level was then 'kick-started' when the UK Department of the Environment seconded an official to DG XI to expedite the drafting of the Directive. The newly-adopted OECD recommendation gave added impetus to the calls for EC action, and also provided a new name for the draft Directive – IPPC rather than 'integrated permitting'.

Preparation of the proposed Directive proceeded through many drafts and took from around mid-1991 to 14 September 1993, when a formal Commission proposal was finally tabled.[20] Negotiation of the Directive took another three years, leading to its adoption on 24 September 1996. During this process of drafting and negotiation, many elements of the text were modified, some provisions were deleted, and others added. Some of the changes which occurred as the Directive emerged are noted in the summary of its final form set out below.

IV The Final Product: A Summary of the Directive

1 *Purpose*

The purpose of the Directive is the achievement of integrated prevention and control of pollution from certain industrial activities, in order to attain 'a high level of protection for the environment taken as a whole'.[21] This is to be accomplished by preventing or reducing emissions to air, water and land, including through measures concerning waste. In practice, a system of permitting is to be implemented.

---

18 Appendix, paragraph 6.

19 Royal Commission on Environmental Pollution, 5th report, *Air Pollution Control* (1976), HMSO, London.

20 Proposal for a Council Directive on integrated pollution prevention and control, COM (93) 423, OJ C 311, 17 November 1993.

21 Art. 1.

The Directive differs from previous EC legislation which dealt separately with emissions to air, emissions to water and the management of waste. In theory, manufacturers and authorities thus should have to think about all emissions and impacts in the design of the whole plant as 'clean technology', rather than relying on 'end-of-pipe' techniques.

The Directive may also be thought of as a 'framework' measure, since it provides for common EC emission limits to be adopted at a later date, and creates a new structure within which certain existing quantitative EC standards are to be applied.

2 *Scope*

The Directive applies to six categories of industry: energy; production and processing of metals; minerals; chemicals; waste management; and 'other'. The 'other' group includes facilities operating in the areas of pulp and paper production, textile treatment, tanning, food production, and the intensive rearing of poultry and pigs. Within each category, the scope is defined further either by relation to the nature of the process or product (e.g. refining of oil) or the size of the operation (e.g. production of ferrous metal above 20 tonnes per day).[22]

The scope of the IPPC regime changed significantly during negotiations as the shape of the Directive materialised. Some examples of the more important modifications are as follows:

- A provision was added during negotiations which excludes application of the IPPC regime to installations used for 'research, development and testing of new products and processes'. Given the polluting potential of some such installations, this may prove to be a serious omission.
- In other areas, however, the Directive's scope was increased. For example, landfills were added to the list of installations to be covered by the IPPC regime, as were large installations for the intensive rearing of poultry and pigs. In the latter cases, however, the Directive states that the setting of emission limit values is to take account of practical considerations,[23] which raises the question of whether practical considerations are to be ignored in authorising other types of installations.
- In some cases the text was made more ambiguous, and it is likely that this was intentional in order to give Member States some discretion in choosing where (and where not) to apply the Directive. For the chemicals sector, the scope of the Directive is defined not by reference to clear boundaries or activities but by stating that coverage is to apply to installations for the production of chemicals 'such as' those presented in a list of example substances. The extent to which Member States adopt similar interpretations of this provision in applying the Directive to their chemical industry sectors remains to be seen.

---

22 Annex I.
23 Art. 9(3).

3 Timetable for Implementation of the Directive

The Directive creates a requirement for industrial facilities which fall under its scope of coverage to be made subject to authorisation through permitting. Following the Directive's adoption on 24 September 1996, it was published in the *Official Journal of the European Communities* on 10 October 1996[24] and entered into force on 30 October 1996.[25] The requirement to apply its authorisation requirements to new plants must be given effect no later than three years after the Directive's entry into force, creating a deadline of 30 October 1999.[26] The majority of the Directive's provisions will have to be applied to existing plants within eight years thereafter.[27] The decision as to whether a facility is 'new' or 'existing', when it has already been built but has not yet come into operation or been authorised under a pre-IPPC regime, depends on whether it has been the subject of a 'full request for authorisation' in the view of the competent authority.[28]

In addition to setting deadlines for application of the IPPC regime to industrial installations, dates corresponding to various other activities are also specified. A summary of key dates under the Directive is presented in Figure 1.

24 OJ L 257, 10 October 1996.
25 Art. 22.
26 Art. 21(1).
27 Art. 5(1).
28 Art. 2(4).

Figure 1. Summary of Key Dates under the IPPC Directive

	Legislative Passage	Industrial Application	Information Exchange	Reporting on Implementation
1996	Directive adopted on 24.9 Directive published on 10.10 Directive in force on 30.10			
1997 1998				
1999	Directive in effect by 30.10	Application to new plants from 30.10		Questionnaire on implementation to Member States by 30.4 (and every 3 years thereafter)
2000				
2001			Member State submissions on emission limits and BAT assessments by 30.4 (and every 3 years thereafter)*	
2002				
2003				National reports to Commission (the first covering 30.10.99 – 29.10.2002) by 29.7 (and every 3 years thereafter)

(continued)

	Legislative Passage	Industrial Application	Information Exchange	Reporting on Implementation
2004				Commission implementation report by 29.4 (and every 3 years thereafter)
2005 2006				
2007		Application to existing plants by 30.10 (and repeal of Directive 84/360/EEC)		

* The Commission is also to publish the results of information exchanges every three years, but no date is set for the first such publication.

4 *Transitional Provisions*

Because the IPPC Directive is intended to replace the previous EC approach towards pollution control which has operated along medium-specific lines, its provisions include several modifications to the body of old legislation. In order to effect an organised transition, the requirements of Directive 76/464/EEC on discharges of dangerous substances to water[29] and Directive 84/360/EEC on emissions to air from industrial plants[30] will continue to apply to facilities until they have been brought under the regime of IPPC.[31] Directive 84/360/EEC, the coverage of which is entirely subsumed by IPPC, will be repealed after the IPPC Directive has been in force for 11 years (30 October 2007). Directive 76/464/EEC applies to some facilities which are beyond the scope of IPPC and so will remain in force in respect of those facilities. Various other measures affected by the introduction of IPPC are to be considered for possible amendment before the repeal of Directive 84/360.[32]

5 *Best Available Technology*

The basic technology requirement of the IPPC Directive is 'Best Available Techniques' (BAT). At first sight, this looks more demanding than the standard of 'Best Available Techniques Not Entailing Excessive Costs' (BATNEEC) applied in previous EC legislation, for example through Directive 84/360/EEC on industrial air pollution which IPPC will replace. However, 'available' is defined to mean economically and technically viable, taking into consideration costs and advantages,[33] suggesting that the overall meaning of BAT is closer to BATNEEC than it might first appear. In determining BAT, special consideration must be given to certain factors, including: the use of low-waste technology; the use of less hazardous substances; the furthering of recovery and recycling; the consumption of raw materials and water; and energy efficiency.[34] An obligation is placed on Member States to ensure that their competent authorities remain informed of BAT developments.[35]

6 *Basic Obligations Imposed on Operators*

The Directive provides that Member States' competent authorities must ensure that installations are operated in such a way that certain general principles are followed. These are: to take all appropriate preventive measures against pollution, in particular through the application of BAT; to ensure no significant pollution is caused; to avoid

---

29 OJ L 129, 18 May 1976.
30 OJ L 188, 16 July 1984.
31 Art. 20(1).
32 Art. 20(3) and Annex II.
33 Art. 2(11).
34 Annex IV.
35 Art. 11.

waste production and recover or safely dispose of waste produced; to use energy efficiently; to take the necessary measures to prevent accidents; and to protect and clean up the site upon cessation of the industrial activity.[36] This article was added at a late stage in the development of the Directive and is generally to be welcomed. The principles enunciated all seem desirable from an environmental point of view, and complement the objective of minimising emissions in order to achieve a high level of protection for the environment as a whole. However, two problems arise in interpreting the article, as outlined below.

The requirement for installations to be operated following the basic principles summarised above could be read as imposing a continuous obligation for operators to satisfy the principles, regardless of any conditions imposed in their operating permits. In other words, if an operator were to become aware that further measures could be taken to prevent pollution in accordance with BAT, an obligation to implement those measures would arise independently of any permit conditions or instructions from the competent authority. However, the Directive appears to negate this interpretation by stating that 'it shall be sufficient if Member States ensure that the competent authorities take account of the general principles ... *when they determine the conditions of the permit*'[37] (emphasis added). This could be interpreted as meaning that operators do not have any residual obligations beyond compliance with their permit conditions. If this is so, it must be seen as a weak point in the Directive, in part because it makes the success of the IPPC regime almost entirely dependent on the effectiveness of the competent authorities, and also because it maintains static performance requirements between permit reviews rather than demanding continuous improvement. The problem is exacerbated because, as is discussed later, permits are only to be reviewed 'periodically'[38] which could in practice leave permits unmodified for many years.

A further problem is that the status of the general principles is confused by apparently contradictory text. In the first paragraph it is stated that 'Member States *shall* take the necessary measures to ensure that installations are operated in such a way that' the general principles are followed (emphasis added). In contrast, the last paragraph of the article allows that 'it shall be sufficient if ... competent authorities *take account* of the general principles' (emphasis added). Again, therefore, the principles appear to have been devalued.

7 *Permitting Requirements*

The Directive requires Member States to ensure that installations covered by the IPPC regime are not operated without a permit.[39] A 'permit' is defined as 'that part of the whole of a written decision (or several such decisions) granting authorisation to operate all or part of an installation, subject to certain conditions which guarantee that the

36 Art. 3.
37 Art. 3.
38 Art. 13(1).
39 Art. 4.

installation complies with the requirements of this Directive'.[40] It is clear from this definition that Member States need not operate a system which grants a *single* permit for each site covered by the IPPC regime.

It had in fact become clear during the early stages of drafting that agreement on the requirement for a single permit would not be achievable, in view of opposition from some Member States where administrative arrangements provide multiple permits for individual plants. As an alternative route to integrated permitting, the Commission had pursued the concept of a 'supreme competent authority' which would coordinate the licensing procedure where more than one regulator was involved. This idea was included in the original proposal but did not make it into the final Directive, again due to opposition from Member States with several authorising authorities who disliked the idea of making one body 'supreme' over the others.

The compromise provision included in the Directive, which seeks to ensure at least some integration in permitting, is that 'the conditions of, and procedure for the grant of, the permit are [to be] fully coordinated where more than one competent authority is involved, in order to guarantee an effective integrated approach by all authorities'.[41] The effectiveness of the Directive in achieving improved control relative to single-medium regulation is likely to turn to a significant extent on the way in which this provision is respected and enforced. There is a very real danger that, if proper weight is not given to its requirements, the objective of applying a holistic approach to industrial installations will be lost and the 'coordinated permit' may become little more than an assembly of medium-specific conditions attached physically together.

Operators must be required to advise the competent authorities of any changes in their operations, and any substantial modifications must be made subject to prior authorisation.[42] Competent authorities must reconsider and, if necessary, update permit conditions 'periodically'.[43] The Directive does not specify a particular interval between permit reviews, leaving this to the discretion of individual Member States. In contrast, the Commission's original proposal had specified mandatory review of permits every 10 years – which still appears a long time compared with the pace of technological (and hence BAT) developments. In the Directive as adopted, mandatory reconsideration must be undertaken only when certain criteria are met, *inter alia* when excessive pollution occurs or when developments in BAT allow significant emission reductions without excessive cost.[44] There may be wide variation in the interpretation of when these conditions arise among the Member States.

Applications for permits are to describe: the installation and its activities; the materials, substances and energy used or generated; sources of emissions; site conditions; emissions from the installation and significant environmental effects; proposed techniques to prevent and reduce emissions; measures for the prevention and

---

40 Art. 2(9).
41 Art. 7.
42 Art. 12.
43 Art. 13(1).
44 Art. 13(2).

recovery of waste; further measures to comply with the basic obligations of operators; and proposed monitoring measures.[45] Permit applications must include a non-technical summary. Information generated in respect of any other legislation, including Directive 85/337/EEC on Environmental Impact Assessment (EIA)[46] and Directive 85/201/EEC on the control of major accident hazards,[47] may be included as part of an IPPC application.[48]

A permit may not be issued unless it can be guaranteed that an installation will meet the requirements of the Directive.[49] The original proposal had contained a time limit of six months for taking decisions on permits, but this had been deleted by the time the Directive was adopted. In the case of new installations or substantial modifications, certain information obtained during the EIA under Directive 85/337/EEC must be taken into account in granting a permit.[50]

There are some specific requirements for the content of permits. All permits must include details of the arrangements made for air, water and land,[51] and such measures as are necessary for compliance with the Directive's provisions on general operator obligations and environmental quality standards.[52] Emission limit values or equivalent parameters must be defined for pollutants likely to be emitted in significant quantities, in particular for certain priority pollutants listed in the Directive, and if necessary a permit must prescribe requirements for protection of soil and groundwater and management of waste.[53] In all cases, permits must contain conditions to minimise long-distance and transboundary pollution and to ensure a high level of protection for the environment as a whole.[54] Permits also must contain monitoring requirements and an obligation to provide data to the competent authority for compliance checking,[55] and measures relating to non-normal operations.[56]

8 Emission Limits

Emission limit values or equivalent parameters imposed in permits are to be based on BAT without specifying the actual equipment to be used. However, determination of BAT is to take account of the technical characteristics of the installation, its

45 Art. 6(1).
46 OJ L 175, 5 July 1985.
47 OJ L 230, 5 August 1982.
48 Art. 6(2).
49 Art. 8.
50 Art. 9(2).
51 Art. 8.
52 Art. 9(1).
53 Art. 9(3) and Annex III.
54 Art. 9(4).
55 Art. 9(5).
56 Art. 9(6).

geographical location and local environmental conditions.[57] These factors will vary throughout Europe, as will the consideration of economic factors in the determination of BAT, so it is to be expected that significant differences will emerge in the emission limits applied by the Member States. The Directive recognises this fact, and sets out a procedure for the submission to the Commission of information on national assessments of BAT and emission limits.[58] This provides the basis for the possible establishment of Community emission limit values for the priority substances listed.[59]

The first submission by the Member States on emission limits and BAT assessments is due 18 months after the Directive comes into effect.[60] This date will be 30 April 2001, so it appears that if Community-wide emission limits are to be set under IPPC (which is by no means certain) they are unlikely to be in place before 2002.

In the absence of EC standards established with specific reference to the IPPC Directive, the standards specified in several existing EC instruments are to serve as minimum emission limit values for the relevant facilities controlled through IPPC.[61] The stipulation that these are 'minimum' standards means that if an objective assessment of BAT were to determine a stricter requirement, this should be reflected in the permit. Whether or not this will happen in practice remains to be seen.

In addition, if attainment of any EC environmental quality standard (such as those defined in various Directives for air and water) needs stricter control than is achievable by BAT, then additional measures must be required in the permit.[62] The nature of these 'additional measures' is not defined. However, it is clear that any emission leading to a breach of an environmental quality standard would cause a violation of EC law under the Directive setting the quality standard if not under the IPPC Directive itself.

In the Commission's original proposal for a Directive on IPPC, an article was included which explicitly would have allowed greater emission levels than would have resulted from BAT in areas where environmental quality standards were being met. This was supported by some of the less wealthy Member States who saw it as supporting their economic development objectives. However, it was opposed by the more prosperous nations, and by environmentalists concerned that it would allow near-pristine environments to be polluted up to the level permitted by environmental quality standards rather than seeking to minimise pollution. The clause was removed before the Directive was adopted, although something along similar lines remains in place because of the way BAT is to be applied. More specifically, determination of emission limits is to be based on BAT but is also to take account of the geographical location of the installation and local environmental conditions.[63]

57 Art. 9(4).
58 Art. 16.
59 Art. 18 and Annex III.
60 Art. 16(1).
61 Art. 18(2) and Annex II.
62 Art. 10.
63 Art. 9(4).

9 Derogations

The Directive allows two main derogations from its provisions on individual permitting. Firstly, a facility may be allowed temporarily to exceed emission limits based on BAT or conditions relating to long-distance or transboundary pollution, if an approved rehabilitation plan ensures compliance within six months.[64] Secondly, Member States may prescribe some requirements for certain types of installations in general binding rules instead of including them in individual permits.[65]

Neither of these allowances had been included in the Commission's original proposal. The option of using general binding rules, in particular, needs to be approached with care if the goal of arriving at considered and holistic solutions for each site is not to be missed. This suggests that general binding rules are only likely to be suitable for categories of installations where there is little complexity or variability. It remains to be seen, however, whether such an objective approach will in fact be applied in all Member States. The only explicit requirement stated by the Directive for the use of general binding rules is that they must achieve 'an integrated approach and an equivalent high level of environmental protection as a whole' compared with individual permitting. Inevitably this will be hard to measure, particularly if general binding rules are used to replace individual permitting completely in some sectors, providing no basis for comparison.

10 Enforcement

The Directive includes certain requirements aimed to ensure that the system of IPPC is applied and enforced in practice.[66] Member States must ensure that permit conditions are complied with, and that operators regularly provide competent authorities with results of release monitoring. Operators must inform authorities of any significant incidents or accidents without delay. Operators additionally must provide the authorities with the necessary access and assistance to enable inspections and other monitoring functions to be carried out. However, there is no explicit obligation on operators to report on non-compliance with, for example, emission limits, although such a provision had been included in the Commission's original proposal. Such an obligation could be interpreted to arise under the requirement to report on incidents, but as an 'incident' is not defined this is by no means certain.

---

64 Art. 9(6).
65 Art. 9(8).
66 Art. 14.

11 *Public Information and Participation*

Member States are required to ensure that applications for IPPC permits are made available for public review and comment before a decision is reached.[67] There is no requirement, however, for public comments to be taken into account in reaching the decision. The decision, the permit and any updates thereto also must be publicly accessible,[68] as must the results of any monitoring of releases under the permit.[69] The requirements for public release of data are subject to the restrictions set out in Directive 90/313/EEC on access to environmental information,[70] which means that material may be withheld on grounds of commercial confidentiality or for certain other reasons.[71]

12 *Transboundary Effects*

Where a Member State is aware that the operation of an installation is likely to have significant negative environmental effects in another Member State, or where the Member State which may be affected so requests, the Directive provides for a process of consultation. This requires the details of the permit application to be forwarded to the Member State affected, 'as a basis for any consultations necessary in the framework of the bilateral relations between the two Member States on a reciprocal and equivalent basis.'[72] In other words, there is a legal obligation for the Member State in which pollution originates to provide information to a Member State which would be affected, but any ensuing discussions between the two countries will be outside the scope of the Directive. One country does not have a legal right under the Directive to prevent an authorisation in another. However, the requirement to ensure that 'no significant pollution is caused'[73] applies regardless of the country affected. At least in theory, therefore, any authorisation which would cause significant pollution in another Member State (or even a non-EC country) would be a violation of the Directive.

13 *Inventory of Emissions*

The Directive requires publication by the Commission, every three years, of an inventory of the 'principal emissions and sources responsible' on the basis of data supplied by the Member States.[74] The format for data provision by the Member States is to be determined through a Committee procedure established under the

67 Art. 15(1).
68 Art. 15(1).
69 Art. 15(2).
70 OJ L 258, 23 July 1990.
71 Art. 15(4).
72 Art. 17(1).
73 Art. 3(b).
74 Art. 15(3).

Directive.[75] The same procedure may also lead to proposals for measures to ensure inter-comparability between the data collected under the IPPC regime and other registers.

The topic of the emissions inventory was a source of some debate during the drafting and negotiation of the Directive. At one stage, among some parties at least, it was envisioned as a 'Polluting Emissions Register', similar in scope to the US Toxics Release Inventory. Other parties were opposed to the idea of any form of register or inventory altogether. For its part, the Commission favoured the option of a separate instrument for a rather more detailed register or inventory than was finally provided for, but when it became clear that this was unlikely to be accepted, the compromise text included in the Directive was agreed. Even now the precise nature of the inventory is unclear because the format and particulars of the system have yet to be established.

14 *Exchange of Information*

As described previously, the Directive requires information on national assessments of BAT and emission limits to be submitted to the Commission, as the basis for the possible introduction of common EC standards applicable through the IPPC regime. In addition, the Commission is to organise an exchange of information between Member States and the industries concerned on developments in BAT and associated monitoring.[76] The results of such exchanges are to be published every three years. The Directive does not specify when the first publication is due, however. In fulfilment of this obligation, the Commission is establishing a procedure for the production of BAT Reference documents (BREFs), although what the precise status of these documents will be is not yet known.

15 *Reporting on the Effectiveness of the Directive*

The Directive requires the preparation of reports on its implementation, and its effectiveness compared with other EC environmental instruments.[77] It is not yet clear how a comparison between IPPC and other environmental measures will be made, although it appears certain to demand a large amount of work if it is to yield meaningful results. The implementation reports are to be prepared in accordance with a procedure set out in Directive 91/692/EEC (the standardised reporting Directive).[78] The first report is to cover the first three years after the IPPC regime comes into effect (30 October 1999 to 29 October 2002), and is due to be submitted by the Commission to the Council by 29 April 2004. At that time, some of the answers to the many

75 Art. 19.
76 Art. 16(2).
77 Art. 16(3).
78 OJ L 377, 31 December 1991.

questions which remain concerning the various ambiguities and uncertainties in the Directive may be answered.

V Conclusions

The summary of the Directive above has noted a number of flaws in the EC's IPPC regime, but these must be placed in the proper perspective. Development of any EC law among a large number of Member States is bound to involve a process of negotiation and compromise, and as a result is unlikely to match the clarity of drafting sought in most national legislation. In addition, the current mechanisms which exist for enforcement of EC environmental law are based primarily on determining the adequacy of transposition rather than actual implementation in practice. This means that the effectiveness of the IPPC regime will, to a large extent, depend on the attitude taken by individual Member States towards its implementation. The Directive is, after all, only the broad framework for a licensing regime, not a detailed prescription of every requirement and procedure to be followed. Despite the Directive's ambiguities, therefore, if the Member States have the resolve to introduce an effective and truly integrated system of industrial licensing, they will be able to do so.

Perhaps more important than the precise wording or structure of the IPPC Directive is the fact that it illustrates that integrated – as opposed to medium-specific – environmental thinking has now entered the mainstream of EC environmental policy. The Directive embodies the culmination of a ten-year process of development and negotiation, but at the same time it represents only the first stage of a more significant journey. The Directive's application of an integrated approach to industrial permitting is just part of a much broader IPPC concept enunciated by the OECD and others. Yet much – indeed probably most – of our current environmental policy and legislation retains a single-medium focus rather than reflecting the philosophy of integration. Clearly, there is still a long way to go.

The Relationship between the IPPC Directive and Other EC Environmental Law

*J.H. Jans**

I Introduction

In this chapter I want to make some observations from a more general perspective. How does this Directive relate to other EC Environmental Law? How does this Directive contribute to the development of EC Environmental Law? Are there any, again looking from a more general perspective, interesting new features in this Directive, worthwhile being pursued in other Directives also? The contrary will also be addressed: which elements are not or not sufficiently included in the Directive but should have been?

II The Integrative Character of the IPPC Directive

If one looks to European Environmental Legislation – I refer here to all the roughly 200 Directives and Regulations of the Commission and Council relating to environmental protection[1] – one comes very rapidly to the conclusion that the bulk of EC environmental legislation is sector related. We have Directives on water pollution, air pollution, chemical substances, noise, waste, nuclear safety and nature protection.

In general, one can say that an integral approach to combating pollution is very seldom indeed. It is very rare to find in a Directive on water pollution a reference to a Directive on waste, or on air pollution. But even within one sector the approach can be fragmented by reason of the existence of several different Directives, dealing with one sector-related type of pollution. Take for instance the area of water pollution.

---

* Jan H. Jans is Professor of European Union Law at the Europa Institute, Faculty of Law, University of Amsterdam.

1 See for an overview Chapter 8 of my *European Environmental Law*, European Monographs No. 12, Kluwer Law International (1995).

European environmental legislation designed to limit the pollution of both freshwater and seawater is very fragmented. First of all there are Directives setting out quality objectives. These establish minimum quality levels for freshwater and seawater, based on the different potential uses of water. Standards are laid down which the water in question has to meet. Examples are Directives establishing quality standards for surface water intended for the abstraction of drinking water,[2] for water intended for human consumption,[3] bathing water,[4] fishing waters[5] and shellfish waters.[6] Secondly, there are Directives relating to the protection of the aquatic environment against pollution by dangerous substances.[7] Specific regulation can be found in the Directive on urban waste water,[8] or the Nitrate Directive[9] aiming to reduce water pollution caused or induced by nitrates from agricultural sources. Even a particular type of industry causing specific problems can be the subject of a Directive. I refer of course to the Titanium Dioxide Directive.[10]

Indeed, not very integrated at all. 'Horizontal environmental legislation' by which I mean legislation which, in contrast with sectoral legislation, relates not only to a single sphere of the environment (water, air, noise, soil, radiation etc.) but which examines a given activity from a variety of environmental viewpoints is still very rare at the European level. An example of such legislation is first of all the well-known Environmental Impact Assessment Directive.[11] This Directive makes it possible to assess all kinds of decisions in terms of their potential effects on the environment and on natural resources. Secondly, at least to a certain extent, one could also point to Directive 82/501 on the major-accident hazards of certain industrial activities, better known as the Post-Seveso Directive.[12] It provides for measures to protect the public and the environment from the consequences of possible major industrial accidents. Other horizontal measures include the Directive on the freedom of access to information on the environment,[13] Regulation 880/92 on a Community eco-label award scheme[14] and the Eco-Audit Regulation.[15] And that is more or less it.

And now we have the IPPC Directive. Of course one can criticize various aspects concerning the integrated character of the Directive very easily. It does not achieve any

2 OJ 1974 L 194/34, later amended.
3 OJ 1980 L 229/11, later amended.
4 OJ 1976 L 31/1, later amended.
5 OJ 1978 L 222/1, later amended.
6 OJ 1979 L 281/47, later amended.
7 OJ 1976 L 129/23, later amended.
8 OJ 1991 L 135/40, later amended.
9 OJ 1991 L 375/1.
10 OJ 1992 L 409/11.
11 OJ 1985 L 175/40, later amended.
12 OJ 1982 L 230/1, later amended.
13 1990 L 158/56.
14 OJ 1992 L 99/1.
15 OJ 1993 L 168/1.

integration with respect to those activities not listed in Annex I of the Directive. One could indeed question whether waste management is indeed sufficiently included in the system of nature protection requirements. One could also question whether the coordination requirement of Article 6 is indeed strong enough to ensure that the procedure and conditions for the granting of the permit are fully coordinated where more than one competent authority is involved. But irrespective of this, it is my opinion looking at the Directive from a more distant and general point of view, that we must welcome it wholeheartedly as setting a major step forward in removing at least some of the watershed between the different environmental sectors.

III Other General Observations

Although a new step in the direction of a more integrated approach, this IPPC Directive is, in other aspects, rather traditional. It is of course a traditional command and control Directive, with rules, permits, etc. There is hardly any attention to more alternative instruments, like covenants or a market-oriented approach. In this respect the Directive may be a little disappointing and a more creative approach would have brought us more interesting results. The promises made in the Fifth Action Programme[16] in this respect are not being met in full.

IV Some Interesting Features of the Directive From a More General Point of View

After these more general observations I now want to highlight some of the, in my view, more interesting peculiarities of this Directive.

First of all, I want to point to the way in which the general principles governing the basic obligations of the operator can be implemented by the Member States. The final sentence of that provision reads that 'it shall be sufficient for the purposes of compliance with this article, if Member States ensure that the competent authorities take account of the general principles set out above when they determine the conditions of the permit'. I want to make the following remarks in this respect. First of all, I have to point to Article 8(6) of the Directive making it clear that *certain* permit-requirements can be replaced, for *certain categories of installations* by a set of generally binding rules. It is therefore excluded, in my view, that for Annex I activities the permit-requirement can be replaced completely. Only partial replacement seems to be allowed under this provision.

Secondly, I wish to make the following remark. Normally Member States are required to adopt laws, regulations or binding administrative measures to implement the Directive. Article 20 of the Directive states this quite clearly. Also one can point

16 OJ 1993 C 138/1.

at the abundant case law of the Court of Justice in this respect. The German TA Luft cases and the French Air Quality cases are examples of this.[17] Legal certainty requires implementation by law. In essence, the case law of the Court means that a Directive has to be transposed into binding provisions of national law, in a manner which fully satisfies the requirements of clarity and legal certainty, and affords interested parties recourse to the courts. From this case law it is also quite clear that mere compliance with the provisions of the Directive in granting permits is not a sufficient means of implementation. Article 2a of the Directive, however, seems to derogate from this case law. As long as the competent authorities take into account the general principles, this provision is complied with. A possible consequence of this is that if the operator violates these basic principles one cannot say that there is a violation of the Directive. Possibly a violation of the permit conditions and national law, but not of EC law. I am therefore not very happy with this provision, because it makes the supervision of *factual* compliance and enforcement of the Directive in the Member States a very difficult task.

Talking about compliance, I now have to point to Article 13. Member States shall take the necessary measures to ensure that, *inter alia*, the conditions of the permit are complied with by the operator when operating the installation. This is not an unusual provision. Enforcement of Community environmental law is indeed primarily the responsibility of the Member States. Article 130s(4) of the Treaty states in so many words that the Member States shall implement the environment policy. The Court has held in connection with Article 5 of the Treaty that, to the extent that environmental Directives and Regulations do not provide for specific enforcement obligations, national enforcement measures must be proportionate, effective, preventive and non-discriminatory. Furthermore, national rules dealing with infringements of Community environmental standards should not be less favourable than those governing infringements of national environment rules.

So what has to happen if the operator disregards the conditions of the permit? Article 13 indeed seems to indicate that there is ample discretion for the Member States, at least in respect of the question *how* standards contained in environmental Directives are enforced. Of course this is within the limits of Article 5 of the Treaty (effective, preventive and proportional). As a result of this discretion, an individual does not have a right (i.e. there is no direct effect) to demand *specific* enforcement action, unless this has been expressly provided for in the Directive itself. If we look at some of the other environmental Directives which do contain specific enforcement actions, Article 13 is not a very exemplary provision. An example of such a provision is for instance Article 5(4) of the Water Framework Directive 76/464.[18] This Directive

---

17 Case C–361/88 *Commission* v. *Germany* [1991] ECR I–2567; case C–59/89 *Commission* v. *Germany* [1991] ECR I–2607; case C–13/90 *Commission* v. *France* [1991] ECR I–4327; case C–14/90 *Commission* v. *France* [1991] ECR I–4331; case C–64/90 *Commission* v. *France* [1991] ECR I–4335.

18 OJ 1976 L 129/23, later amended.

requires national authorities to establish emission standards in authorizations to prevent water pollution. The provision states:

> Should the emission standards not be complied with, the competent authority in the Member States concerned shall take all appropriate steps to ensure that the conditions of authorization are fulfilled and, if necessary, that the discharge is prohibited.

This provision implies that it is not permitted to tolerate discharges which violate the Community emission standards contained in the Directive. Some other environmental Directives similarly contain enforcement provisions. Take for instance Directive 94/67 on the incineration of hazardous waste.[19] Directive 94/67 requires the setting and maintaining of appropriate operating conditions and emission limit values for hazardous waste incineration plants. Plants for the incineration of hazardous waste shall be operated in order to achieve a level of incineration as complete as possible (Art. 6(1)) and in such a way that at least certain emission limit values are not exceeded (Art. 7). The Directive requires high-standard measurement techniques to monitor emissions to ensure compliance with the emission limit values (Art. 11). Should the measurements taken show that the emission limit values have been exceeded, the plant concerned shall not continue to feed hazardous waste while failing to comply with these limit values (Art. 12(1)).

It is not completely clear how these enforcement provisions in specific Directives relate to Article 13. To give an example: an IPPC Annex I activity violates emission standards taken in the framework of Directive 76/464 or Directive 94/67. Are compliance measures to be taken according to Article 13 of the IPPC Directive leaving much discretion to the Member State or according to the much more strictly drafted Water Framework Directive or the Waste Incineration Directive? The transitional provisions of Article 19 of the IPPC Directive seems to indicate that once an installation has an IPC permit the enforcement provisions of Article 5 of Directive 76/464 and Directive 94/67 (one of the Directives mentioned in Annex II) are no longer applicable. Article 19 of the IPPC Directive is rather explicit in stating that Article 5 of the Water Framework Directive will not be applicable. With respect to the Annex II Directives Article 19 states that 'the relevant provisions concerning authorization systems' will, in due time, no longer be applicable. Are the compliance provisions of the Waste Incineration Directive to be regarded as such? Probably so, but I am not quite certain.

But if this interpretation is correct then the IPPC Directive has to be regarded, from an enforcement and compliance point of view, as a drawback from the existing legal situation. Even more paradoxical is that for those discharges of substances into the aquatic environment which do not fall under the IPPC Directive (notably the minor damaging activities), the more stringent and strict enforcement rules must be implied.

---

19 OJ 1994 L 365/34.

The second interesting feature, again looking from a more general point of view, is the extensive and sometimes detailed rules and requirements regarding procedural aspects. These relate to the rules on applications for permits, rules to coordinate multiple national licensing authorities, rules on reconsideration and updating of permit conditions, as well as access for the public to applications for permits. In this age of subsidiarity it is refreshing to see that it is still possible to produce harmonisation also in this area of environmental law. This Directive, in my opinion, produces a fine balance between harmonization of substantive and formal and procedural environmental law. It is therefore quite clear that any further legislative activity on the part of the Council in harmonizing other aspects of procedural environmental law does not meet any legal obstacles. The Council is indeed quite competent, on the basis of Article 130s(1) to decide on these types of issues also. Maybe it will be possible in the near or distant future to include in this Directive provisions on the access of the public to the integrated permit procedure. The only current provision in this respect is Article 14(1) where the public has been granted the right to comment on an application for a permit. More elaborate and detailed provisions, on hearings for example, could have been useful.

Provisions relating to access to national court proceedings where the standards of the Directive have been violated are also lacking. A fully harmonized and integrated Directive on pollution prevention should have addressed the problem of access to court proceedings.

In relation to this one can also point to the provisions of the Directive on access to information.[20] In addition to the well-known Directive on the freedom of access to information on the environment, Member States have to ensure that applications for permits for new installations or for substantial changes are made available to the public for an appropriate period of time. This is indeed an extension of the rights which the public have under the Freedom of Information Directive. On the basis of that Directive the public have only a passive right to obtain a copy of the application; there is no active duty for the competent public authority on the basis of that Directive to publish it. Furthermore, the decision of the public authority must also be made available to the public.

Particularly interesting, although I am not quite sure what the precise meaning is, is paragraph 2 of Article 14. The results of emission monitoring as required under the permit conditions and kept by the competent public authority must also be made available to the public. In principle, emission monitoring data kept by public authorities would also fall under the scope of the Freedom of Information Directive and must be made public upon request. However, the public authority could refuse access to that information if it is regarded as commercially sensitive information, according to Article 3 of the Freedom of Information Directive. The exceptions mentioned in Article 3 of that Directive are also applicable to the information to be provided under the IPPC Directive. However, in the Commission's proposal it was stated that these restrictions

20 OJ 1990 L 158/56.

on public access to the information gathered during monitoring may be applied only in exceptional circumstances. This clause has been deleted by the Council in the final text.

This seems to indicate that the Member States have recourse to the normal restrictions on the freedom of information as regards emission monitoring data. But if that is the case, why is it necessary to mention explicitly emission monitoring data being made available to the public, if this data anyway falls within the scope of the Freedom of Information Directive? One could therefore argue that with respect of the results of emission monitoring requirements the IPPC Directive provides more or less a presumption of openness. In providing explicitly that this information has to be made available to the public, the text demonstrates that normally this information cannot be held secret. My conclusion is therefore that although the Council has deleted the clause in the Commission proposal which only restricts the openness of emission monitoring data in exceptional circumstances, this restriction is still implicit in the text of Article 14 of the IPPC Directive.

The final element to which I wish to draw attention to are the rules in Article 16 on transboundary effects and in particular paragraph 2 where it is stated that applications for permits should also be made available to the public of the Member State likely to be affected by the planned activity. These non-inhabitants will also have the right to comment on them before the competent authority reaches its decision. The final version of this article is much better drafted compared with the original proposal of the Commission. This original proposal only contained a reference to the procedure to be followed in the case of Environmental Impact Assessment, according to the provisions of the EIA Directive.[21]

As you may know, I have had a keen interest in the subject of the legal protection of citizens in cases of transboundary environmental pollution for a number of years.[22] Finally, in the Directive we can find particular attention to the legal position of non-inhabitants. They will have the right to comment on the application. However, I am by nature very suspicious. And here I find just one word missing. It is not enough that they have the right to comment, they should also have the *same* right to comment as the nationals of the state where the activity is being carried out. But maybe I am just too suspicious and the omittance of the term 'same' has no practical relevance.

Also peculiar is that the right comment on the application, as well as the right to be informed, is only attributed to the public of the Member States. Why is that? Is there not a need to inform the Swiss public or the Czech public if in their border area activities take place which have significant negative effects as regards their environment? I see no legal reason whatsoever to restrict these rights to the Member States

21 OJ 1985 L 175/40, later amended.

22 See my doctoral thesis entitled *Grensoverschrijdend milieurecht*, Groningen (1987), and the German translation thereof: *Grenzüberschreitendes Umweltrecht – im niederländisch/deutschen Grenzgebiet: europäisches Gemeinschaftsrecht, Völkerrecht, niederländisches und deutsches Verwaltungsrecht* (1990).

of the European Union. And it is certainly not in conformity with the text and spirit of Principle 10 of the Rio Declaration on Environment and Development in particular.

V Concluding Remarks

To what conclusion does this lead? Well, despite all the critical remarks I have just made, this IPPC Directive must be welcomed. It is an important step in the development of EC Environmental Law. You could say: European Environmental Law is finally growing up and has become more mature. I hope that in the future this process may result in a further internal harmonization and integration of this most interesting and important area of European policy and law.

The IPPC Directive in National Law Context

Integrated Prevention and Pollution Control: the UK Experience

R. Macrory[*]

I Introduction

EC environmental policy-making is often as much a bottom-up approach as it is a top-down one in that initiatives developed in one Member State may gradually work its way upwards to form the nucleus of Community legislation. Formal legislation on integrated pollution control was introduced in the United Kingdom in 1990 under Part I of the Environmental Protection Act 1990, though its genesis goes back considerably further, and it is well known that British experience had significant influence on the design of the IPPC Directive. The purpose of this Chapter is to examine aspects of the British system and some of the legal and practical challenges being encountered in the effort to translate the concept of integrated controls into operational reality since this may shed some light on the implications of the Community proposals. Many of the provisions in the proposed Directive would resonate with a British environmental lawyer, though the Directive is still likely to require changes to the British legislation, and it is perhaps significant that Community law uses the term 'Integrated Prevention and Pollution Control' (IPC) while the British system omits the word 'Prevention' and confines itself to 'Integrated Pollution Control'. This particular difference, however, may be more symbolic than of real importance and it will be seen that prevention does play an important role in the British controls. It is not, however, my intention to conduct a line by line comparison of the British legal provisions and the proposed Directive since this is best left eventually for detailed discussion between British lawyers, the Government, and the European Commission. But it is worth bearing in mind that the concept of integration in the context of pollution control can imply quite distinct elements.[1] At one level it may encompass improved administrative coordination

* Barrister, Professor of Environmental Law, Imperial College London.

1 See further R. Macrory, 'The Scope of Environmental Law', in: G. Winter (ed.), *European Environmental Law*, Dartmouth, Aldershot (1996).

between regulatory bodies dealing with pollution control or the reorganization of such bodies into signal agencies.[2] It may also imply strengthening the linkages between pollution control legislation and other related systems of control as happened in the United Kingdom in 1988 with the introduction of environmental assessment procedures, largely located within a land-use development control system.[3] Or, from a different perspective, it may imply deeper comparisons between the impacts of pollutants on different environmental media of air, water and land. This latter notion is very much at the heart of the new British controls, and has proved an intellectual and technical challenge. In contrast, it is arguable that what is being offered by the present Community proposal is essentially coordination requirements at a fairly superficial level – whether this will eventually and inevitably lead to more profound integration must still remain questionable.

To understand how the current British system of integrated pollution control came into being it is necessary to step back a few years, say to the mid-1970s. If one examined the structure of environmental law as it then existed, the basic position was one that was familiar in many countries. There was separate specialized legislation regulating discharges into the air, water, and waste on land, all essentially based on traditional licensing systems. Different authorities were responsible for granting licences for discharges into each medium, and operated at different hierarchial levels. Emissions into the atmosphere for most large-scale and potentially problematic industries were regulated by a small, specialized Inspectorate within a Central Government Ministry, the Department of the Environment, employing highly qualified technical staff, using a lot of professional judgment and discretion, but criticized as having less concern with public relations or public involvement in decision-making. Although the legislative base and scope of their controls changed over the years, the fundamental structure had remained constant since the mid-nineteenth century.[4] Discharges into surface and groundwaters were regulated by non-elected regional water authorities, but with greater requirements of public involvement and participation in decision-making being introduced by legislation in 1974. In 1989, following the privatization of the utility functions of these authorities as they related to water supply and the provision of sewerage services, the regulatory functions were subsumed into a single body, the National Rivers Authority, again a non-elected body operating at arms's length from Central Government.[5] The regulation of the disposal of waste on land was the responsibility of local authorities, generally at county council level, employing

---

2 For example, the creation of the French environmental agency, the Agence de l'Environnement et de la Maitrise de l'Energie in 1990, the Irish Environmental Protection Agency in 1990, and the Environment Agency in England and Wales, together with the Scottish Environment Protection Agency, in 1996.

3 Town and Country Planning (Assessment of Environmental Effects) Regulations 1988 SI 1988/1199 and associated regulations.

4 See generally for the history of the controls Ashby and Anderson, *The Politics of Clean Air*, Clarendon Press, Oxford (1981).

5 Described in administrative terms as a non-departmental public body.

specialized legislation introduced under Part II of the Control of Pollution Act 1974. There were few national emission, imission, or environmental quality standards contained in legislation which constricted the discretion of regulating authorities, and the conceptual approach towards pollution control was quite different for each medium. For the atmospheric emissions, the Inspectorate operated what was essentially a technology-based approach, though perhaps not one that would be recognized as such in other countries. The key legal criterion was that the operator had to use the 'Best Practicable Means' to prevent or reduce emissions, but as the definition sections in the Draft Directive illustrates, each of these words is riddled with difficult value and non-technological concepts. Yet it was a very closed system of regulation with little resort to legal challenges; indeed, the key criterion, 'Best Practicable Means', had been contained in the legislation since 1874, and there had not been a single case in the higher courts providing judicial interpretation of the phrase in that context.

The water authorities in contrast largely based their approach on ambient quality objectives, developed largely at their discretion, and looking at the needs of the receiving waters; working from that basis emission limits on individual discharge consents were established. Waste disposal did not really operate any defined conceptual system, and was heavily criticized in a number of authoritative reports for the great variation in approach between different local authorities.

Finally, and to complete a very important element of the system, there existed – and still does – a complex system of land-use planning consents, which in essence required that proposals for all new buildings or industries have to receive individual planning permission from the local authority with appeal to Central Government. Land use plans provide a presumption but do not guarantee that applications in accordance with plans will be granted, and the receipt of planning permission does not necessarily entitle the applicant to any more specialized discharge consents or licences that may be required.[6] This planning system fulfils a distinctive role in British environmental law. It was until fairly recently the only part of the system where there were extensive rights of public consultation and public participation. The planning laws are deliberately vague and open-ended in defining what the planning decisions were actually about, and were therefore very responsive in allowing changing contemporary concerns at least to be echoed during the decision-making process. Thus we have seen planning controls move away from being concerned with purely local impacts on amenity to accommodate issues such as alternative locations, the need for development, and the underlying economics. But equally, planning laws are not simply about environmental protection and planning authorities are entitled to take on board factors such as employment and other social considerations. And given the disaggregated nature of the specialized licensing systems, it was perhaps the only stage of the legal process where one could legitimately consider the implications of a development in the round. Finally, unlike operational licence systems which generally build in a degree of

6 Section 54A Town and Country Planning Act 1990 requires that determination of planning applications 'shall be made in accordance with the development unless material considerations indicate otherwise'.

flexibility and the power to change and update conditions, the award of planning permission is in essence a one-off decision granting rights to develop land. Once given, it is difficult to revoke or make subsequent changes to conditions. It was hardly surprising then that when the UK came to implement the 1985 Environmental Assessment Directive it was into this familiar system of land-use planning consents that the requirements of the Directive for most projects were inserted. The system could accommodate these obligations with reasonable ease.

The position then was one of a complex system of controls with different approaches and different authorities. Undoubtedly one of the motivations for IPC and the present Directive was to reduce the administrative burden on industry.

We are still, though, in the 1970s. Enter the Royal Commission on Environmental Pollution, an independent, high-level advisory body entitled to investigate any area of environment pollution and policy it wishes to. At that stage the Commission decided to investigate the relationship between planning controls and air pollution. Potentially overlapping control systems throw up clear legal problems. Could a local authority refuse planning permission because it was worried about pollution impacts and did not trust the ability of the Central Government specialized regulators to control them? Or should one argue that because there was a specialized control system, one should avoid duplication? This was essentially the preferred approach of Central Government but the courts have never been prepared to accept such a black and white line. There have been, for example, a number of well-known instances where the regulators found themselves with a real difficulty in dealing with air pollution problems without closing down the industry and where the main problem was an initial poor siting decision, the prior responsibility of planners.[7]

The Royal Commission explored all these aspects.[8] It did not wish to combine both systems but recognized that each had a very important part to play. The planning system in essence could be a genuinely preventative and precautionary system. Greater consultation between planners and regulators was recommended, as was clearer guidance to planners on the environmental aspects of their responsibilities. The Commission, however, did not favour greater introduction of statutory air quality standards, feeling them to be too constrictive and inefficient, but wanted to develop air quality guidelines to provide a consistent set of basic criteria for planners.[9] But the Commission went further. Although it had initially embarked on looking at such planning and air pollution controls, its inquiry revealed disturbing features about the disaggregated system of pollution controls in general. The Commission became concerned that with different authorities operating their own specialized systems, there was a real danger that when one considered the total load on the environment, sub-

7 Miller and Wood, *Planning and Pollution*, Oxford University Press, Oxford (1983).

8 Royal Commission on Environmental Pollution, 5th Report: *Air Pollution Control: An Integrated Approach*, HMSO, London (1975).

9 The Government accepted this recommendation but shortly afterwards agreed to a number of EC Directives concerning air quality which have led to the introduction of statutory air quality standards in the United Kingdom.

optimal decisions were being taken. The Commission had examples, say, of licence conditions on water pollution which would increase the load of air pollutants. On a larger scale, decisions concerning, say, Flue Gas Desulphurization taken by air pollution regulators have major implications for the disposal of resulting solid waste arising from the process, and the extraction of lime in quarries. In effect, though the term was not in common parlance then, one was searching for an environmental life-cycle analysis to compare different strategies but the legal system did not permit this.

The Commission's key recommendations were two-fold. First, to extend to the remit of the existing Air Pollution Inspectorate to regulate all discharges into whatever media from the processes they covered, and they recommended that it be recalled Her Majesty's Pollution Inspectorate. Secondly, to introduce a new underlying conceptual basis to pollution control – the best practicable environmental option. Essentially this was described as the outcome of a systematic and consultative decision-making process which emphasises the protection and conservation of the environment across land, air and water. The Best Practicable Environmental Option (BPEO) procedure was intended to establish, for a given set of objectives, the option that provides the most benefit or least damage to the environment as a whole at acceptable cost in the long term as well as the short term.

The Government's response was predictably minimalist and resisted any major institutional or legal change.[10] It accepted that the BPEO concept was 'of considerable power and utility', but considered that with better informal consultation and communication across the various regulatory bodies the goals could be achieved. Any reorganization of the regulatory authorities into a single body would 'involve a substantial increase in operational complexity'. In 1986 a push came from the Cabinet Office Efficiency Unit which felt that a combined inspectorate would allocate resources more effectively to areas where pollution problems were most severe. In 1987 the Government made an initial institutional change by combining a number of the existing Government Inspectorates to form a combined Her Majesty's Inspectorate of Pollution (HMIP). But there were no fundamental legal changes to the desaggregated licensing systems, nor did the new agency involve local waste regulatory authorities or the regional water authorities. A year later the Royal Commission returned to the concept of BPEO and published a report which developed the concept further, and advocated major legal changes.[11] Then in 1990 under the Environmental Protection Act major changes were made. A new licencing system for most major industries and based on the concept of integrated pollution control was introduced. The system was operated by HMIP, and regulates a whole list of processes which nearly matches the list in the Directive. Planning controls operated by local authorities remain in place – for a new process planning permission would be required in addition to the IPC licence. The water regulation system remained in place for many types of discharges, and there remain uncomfortable overlaps on a single site where some

10 Department of the Environment, *Air Pollution Control*, HMSO, London (1982).

11 Royal Commission on Environmental Pollution, 12th Report: *Best Practicable Environmental Option*, HMSO, London (1988).

discharges were regulated by HMIP because of their dangerous nature and others by the water regulatory bodies. Disposal of solid waste off-site remained regulated by waste management authorities.

A final piece of the institutional jig-saw has been created under the Environment Act 1995 setting up new Environment Agencies (one for England and Wales, one for Scotland) which came into being on 1 April 1996. The Environment Agency in England and Wales combines the existing functions of HMIP, the National Rivers Authority and the Waste Regulatory authorities, and has proved a difficult administrative exercise to establish.

Despite the new administrative structure, the basic legal regulatory structure has not changed, although the opportunity has been taken to ensure that certain powers concerning enforcement and other matters are at least consistent across the IPC, water pollution and waste management field. Given the major administrative upheaval that has been involved, it was probably politically sensible not to attempt a major change to the pollution legislation, especially as the IPC legislation is still very much at a learning phase in its implementation. Nevertheless, the Royal Commission on Environmental Pollution has recently recommended that in five years time the Government should undertake a major review of the legislation and potential problem areas of ensuring consistency in approach.[12]

As it is, the IPC legislation remains at the heart of the integrated regulatory system, now being operated by the former Inspectorate of Pollution within the new Agency. The controls are based on a licensing system, but the legislation gives extensive powers to the Secretary of State to impose statutory national emission limits, or environmental quality standards with which the Inspectorate must comply, if applicable.[13] It is clear that statutory emission limits or quality standards cannot encompass all the potential substances that may be discharged into the environment, and remain the exception rather than the rule. The legislation therefore imposes explicit general duties on both the regulator and the operator of licenced processes.

The two key duties of the regulators are that in granting a licence and imposing conditions they must:

(a) ensure that the best available techniques not entailing excessive costs are employed to prevent releases[14] into the environment, and where this is not practicable to reduce the release to a minimum and render any residual releases harmless.

(b) where a process is likely to give rise to releases into more than one environmental medium, to ensure that best available techniques are used to minimize pollution of

12 Royal Commission on Environmental Pollution, 19th Report: *Sustainable Use of Soil*, HMSO, London (1996).

13 Environmental Protection Act 1990 s. 3.

14 Environmental Protection Act s. 7. The duty applies to releases of substances prescribed in regulations, but these cover the most serious. With respect to non-prescribed substances, the duty requires the Inspectorate to ensure that the Best Available Techniques Not Entailing Excessive Costs (BATNEEC) is used to render them harmless (i.e. the prevention and minimization duty does not apply to such substances).

the environment as a whole having regard to the best practicable environmental option.

A number of observations can be made about these duties. The legislation has deliberately employed the phrase 'techniques' rather than the more familiar term 'technology' used in Community legislation to emphasise that the Inspectorate is concerned not simply with the machinery of abatement control but also with matters such as operator training and standards and the design and layout of buildings.[15] There is a hierarchy of duties concerning releases, with prevention being primary, followed by minimization and then only rendering harmless any residual emissions (the latter encompassing dispersion and other means of abatement). In practice most industrial works covered by the legislation will involve releases to more than one medium, and the BPEO duty therefore comes into play. This is the first time the concept has appeared in legislative form, but, although clearly derived of the reports of the Royal Commission on Environmental Pollution, it receives no detailed legislative definition. Until the courts are asked to rule on the question, it has been left to the regulator to develop its own methodology for determining the BPEO, an issue discussed further below. Similarly, no detailed legal definition is given of the phrase 'not excessive costs', and currently the regulator is having to develop its own approach to the meaning of the term.[16]

For the operator, apart from the legal obligation to comply with a licence and the conditions attached to it, there is an important residual condition implied into every licence which covers all aspects of the process not covered by any specific licence condition.[17] They must also make use of best available techniques not entailing excessive costs to prevent releases or where this is not practicable to reduce releases and render them harmless. As with the Inspectorate, the operator must employ the BPEO where discharges are to be made into more than one medium.

This residual condition appears to have three key functions. First, it is a fail-safe mechanism which recognizes that it may be impossible for the Inspectorate to impose conditions covering every eventuality. Second, there was concern that by embarking on a rather more formalized process than had hitherto been the tradition in the UK, the result might be unwieldy and legalistic licensing documents. Finally, the duty clearly imposes a considerable responsibility on operators to take a proactive approach towards pollution abatement. They cannot simply rely upon the Inspectorate and assume that

---

15 Environmental Protection Act 1990 s. 7(10): 'References to the best available techniques not entailing excessive costs in relation to a process include (in addition to references to any technical means and technology) references to the number, qualifications, training and supervision of persons employed in the process and the design, construction, lay-out and maintenance of the buildings in which it is carried on'.

16 But the Inspectorate must have regard to Guidance issued by the Secretary of State. In the revised Guidance on IPC issued in 1993, the Secretary of State indicated that for new processes it is the economic condition of the industry sector as a whole rather than the specific company which will determine whether costs are excessive or not.

17 Environmental Protection Act 1990 s. 7(4).

compliance with the conditions themselves was sufficient to absolve them from further responsibilities.

The system has been operating since 1991, it applies to all new processes falling within the regime, and has gradually been extended to a whole range of existing processes. In the context of the present discussion on the IPPC Directive, I would highlight three aspects of the experience to date: (a) the learning curve of operators and the regulators; (b) the relationship between IPC and Environmental Assessment procedures; and (c) the development of methodologies concerning the BPEO.

II The Learning Curve

The new IPC controls are built on a system of regulating emissions into the atmosphere which the regulatory body, Her Majesty's Inspectorate of Pollution and its predecessors, had operated for over 100 years. A distinctive regulatory style has been developed over that period with the inspectors, who generally have at least ten years' industrial experience, resisting a legalist approach in favour of working with industry with a mixture of persuasion, cajoling, and operating on a basis of mutual technical respect. Few prosecutions were brought, and inspectors sometimes described their approach as using an 'iron fist in a velvet glove'.

With the introduction of the new IPC system and the appointment of a new chief inspector, the Inspectorate initially decided to adopt a much more distanced approach. Industry was expected to make licence applications without the benefit of any informal advice or pre-application meetings. This policy lasted for three years, and in many ways proved counter-productive, though it established the fact that this was a new legal regime with a new relationship between regulators and industries.

The application for authorization placed a considerable burden on industry to justify their proposals, including a description of the process, techniques to prevent releases, an assessment of the environmental consequences of releases, and a description of matters on which the applicant would rely to ensure that the operation achieved the objectives of using the best available techniques not entailing excessive costs and, where applicable, the BPEO. The initial applications that were made, especially those relating to the power industry, were largely considered unsatisfactory, and industry had clearly not fully realized the implications of new requirements. One of the senior inspectors commented, 'Her Majesty's Inspectorate of Pollution can be forgiven in some cases for thinking the companies had deliberately set out not to get an authorization. Many companies certainly had not done their homework'.[18] In February 1992 one Inspector told a conference 'one condition of IPC is that we must

18 Quoted in ENDS, *Integrated Pollution Control: The First Three Years*, Environmental Data Services, London (1994).

check the applicant's competence. If he cannot fill in a form properly and supply accurate information on his process can he still be viewed as competent?'[19]

Out of 118 applications that were received during this early period, the Inspectorate served notices requiring more information in all cases. Over half did not consider alternative processes or abatement techniques, and only a quarter gave any consideration to BPEO. From industry's perspective the application forms were unclear and guidance notes unhelpful. Industry and the Inspectorate put down the initial experience to teething troubles, and a rather different approach was introduced with more pre-application meetings and improved forms. Later applications certainly showed improvement but a major independent study[20] in 1993 demonstrated that many applications were still not providing information relating to a full assessment of environmental impacts. Many were more concerned with human impacts rather than the natural environment and for its part the Inspectorate could be criticized for lack of ecological expertise. BATNEEC and BPEO were better addressed, though there was still a lack of clear understanding. To take one example, out of 60 applications for new organic and inorganic chemical processes, 27 gave serious consideration to alternative processes and abatement, while 12 addressed BPEO fully. The Inspectorate did not appear to find it easy to persuade them to address these issues more fully.

III Relationship with Environmental Assessment

The Inspectorate required that applications for IPC licences contained a full assessment of the total environmental impacts of releases caused by the new processes. However, as previously indicated, the requirements concerning environmental assessment under the 1985 Directive were largely integrated into our land-use planning system. Environmental assessment therefore comes come into play where new projects or substantial alterations to existing premises requiring planning permission are proposed, and the procedures are handled by local authorities. Clearly the IPC regime does not take on all the requirements of environmental assessment, and is essentially orientated around processes and emissions. The visual impact of a project, for example, would fall outside the remit of IPC.

In practice, an industry proposing a new project would generally first secure planning permission, and only then would it seek the necessary operating licences, including IPC. The Inspectorate (now the Environment Agency) would normally be requested to consult, but in considering whether a proposed project may have a significant impact on the environment (for the purposes of Annex II projects) a local planning authority is under no duty to consult HMIP. Generally, the Inspectorate found the information contained in environmental statements prepared by applicants for planning permission less valuable for their own needs than they had hoped.

19 Ibid.

20 The ENDS Report, see *supra* note 18.

In 1993 the United Kingdom Environmental Law Association and Institute of Environmental Assessment, an association established to improve professional standards in the field, conducted a joint study of the linkages between the environmental assessment and IPC regimes. The study recognized that there were many overlaps, but also that it was sometimes impossible to have a single environmental assessment which would fulfil all needs; for example, all operational details of processes were not always known at the initial planning permission stage. Nevertheless, the study concluded that the two systems needed to pay much more regard to each other, and recommended that often the planning process and the IPC process could be considered in tandem, and that it may be possible to employ a single Environemtal Assessment for both. In August 1995, the Inspectorate issued new guidance on their relationship with planning agencies, and indicated that with new industrial processes it would wish to see applications for authorizations prepared at the same time as planning applications. Certainly, with the proposed IPPC Directive it is arguable that closer attention needs to be paid to the overlap with environmental assessment requirements; there are, for example, a number of processes covered by the IPC Directive but not by the EA Directive.

IV BPEO Methodology

Although the term BPEO now appears in the IPC legislation, the phrase is given no statutory definition and as yet there are no court decisions providing guidance. In that context, the Inspectorate has been developing its own initiatives in trying to develop a transparent and rational methodology for determining the BPEO in any particular case. IPC is focused on the releases caused by processes, and as a result the Inspectorate's remit concerning BPEO is rather more restrictive than certainly the Royal Commission on Environmental Pollution imagined when it first developed the concept. In the context of IPC, it would be difficult to employ it to question the need for the process or the products involved, or to conduct a full-life analysis.

One of the difficulties in the first three years of the IPC system was the differing expectations raised by the new legislation. The law provided no definition of what criteria were applicable in determining the BPEO, or how much environmental information would be needed by the Inspectorate to make decisions. Under previous controls, which centred on the equally undefined notion of 'best practicable means', much was left to the professional judgment of the Inspectorate in a fairly informal system which lacked transparency.

Essentially, the IPC process requires both the operator and the Inspectorate to predict the environmental consequences of releases and to weigh them up against each other and in the light of the costs of different control measures. During 1992 and 1993 the Inspectorate began to develop a more formalized methodology which was published in 1994 as a consultation document.[21] The aim was to provide a robust method that was practical and efficient, while recognizing complexities and uncertainties that may inevitably exist.

An initial consultation document on the subject issued in 1993 was very much a first attempt, and was probably over-simplified, being based on the principle that the overall impact of a process could be quantified in numerical terms as a BPEO index. The process option with the lowest value was the best environmental option, and if it was the one chosen by the operator, then there was no problem. If the operator chose an option with a higher value, he would have to justify in detail why this was the practicable option in the circumstances. The fundamental difficulty with this approach was, of course, deriving a workable numerical index. Essentially, the Inspectorate wanted the predicted environmental concentration to be expressed as a proportion of any relevant Environmental Quality Standard or where one did not exist, the Inspectorate's own assessment level. The quotients are then summed up for each medium and then compared with different options.

The approach was criticized as being over-numerical, lending perhaps a spurious air of authority to an exercise which was often putting on similar basis factors which were not comparable. Should discharges to water be treated equally as discharges to air? How should one weigh long-distance releases with local impacts? How could one compare long-term uncertainties (such as global warming impacts) with more immediate effects? These distinctions appeared to become lost in the process. Furthermore, much depended on the ability to derive intolerable and negligible levels, when for many chemicals there was simply a dearth of information. The Inspectorate was aware of this problem, and indicated that where there were no statutory limits, they would look at World Health Organization guidelines, standards produced in other countries, fractions of occupational standards and so on.

The revised consultation document in 1994 incorporates the same basic philosophy but is more refined. It lays less stress on the purely mechanistic approach and states that the final choice must involve weighing all the factors based on expert judgment. In this it reflects the concerns of the Royal Commission on Environmental Pollution in its report on BPEO which emphasised that, 'The presentation of information, opinions and conclusions reached at different stages of the BPEO study is clearly important. The decision-maker, and other persons, should be able to see separately the data or scores relating to each aspect of the decision for each option'.

The approach is still based on the goal of producing an integrated index, based on the ratio of the plant's contribution to Environmental Quality Standards or Environmental Action Levels, but it recognizes that other factors need to be taken on board or may not be readily incorporated into an index – the global warming potential of emissions is given as one example.

21 HMIP, *Environmental, Economic and BPEO Assessment Principles for Integrated Pollution Control*, HMIP, London (1994).

All this information must be initially prepared by the operator, and represents a deliberate policy of forcing operators to think carefully about the environmental consequences of their actions and options. Discussion on the proposed methodology continues with considerable concern expressed by certain sectors of industry that the integrated index is complex and too reliant on number-crunching. In its latest versions the Inspectorate emphasises that the methodology is for guidance only, and that operators may 'use alternative indicators if they are fully justified and provide at least an equivalent degree of transparency and as good an audit trail'.[22]

But the setting of the Environmental Quality Standard or the Inspector's own Action Levels remains critical to the procedure. Standards are the responsibility of law makers but there remain very few in comparison with the total potential. This leaves much to the discretion of the Inspectorate, and the way it exercises its own judgment in deriving Action Levels, or, for example, is prepared to apply a precautionary approach, remains less obvious. With the recent incorporation of the Inspectorate into the new Environment Agency, the Inspectorate will now be working more closely with water pollution and waste management, bringing them in contact with different specialities and administrative cultures, and it remains uncertain as yet to what extent these new administrative structures will affect the development of the BPEO methodology. Certainly, all environmental regulators in the United Kingdom are conscious that they are now operating in a more legally transparent system than was hitherto the case, with applicants (and third parties) more ready to challenge decisions with which they disagree in the courts as legally unsound. While courts in Britain are generally reluctant to interfere with the professional discretion of decision-makers to whom statutory powers have been delegated, they are likely to require that they should be convinced that a rational and consistent approach has been adopted. The Inspectorate's initiatives in developing a BPEO methodology can in part be seen as a response to this new context.

It remains unclear how far the IPPC Directive will eventually involve these concepts of best practicable environmental option, and comparative evaluation. The Directive mentions the protection of the environment as a whole several times and in Article 8 requires the imposition of emission limits to be set for pollutants to prevent the pollution from one medium to another. And the term 'available' clearly includes an economic evaluation similar to that involved in 'practicable'. Nevertheless, it does not address the subject explicitly, and appears to be more focused on improved coordination and cooperation between pollution control authorities. My own view is that true integration implies more than that. For all the criticism that can be made, the attempts being made by UK regulators in developing techniques for comparing options on an integrated basis represent a considerable advance in thinking, and may provide useful lessons for other countries.

22 ENDS Report, *In Search of the Best Practicable BPEO*, Environmental Data Services, London (1995), p. 249.

The IPPC Directive: a German Point of View

*G. Winter**

The political struggles concerning IPPC having ended with the Council approving the Common Position and the European Parliament's consent, an adaptation of the national legal order is now on the agenda. I shall therefore discuss the changes German environmental law will have to undergo. However, as not all of the controversial issues have been settled with sufficient clarity some of them will arise again as questions of interpretation in the transposition stage. To the extent that the Directive has adopted concepts derived from German law their interpretation may be enriched by looking at their doctrinal and practical national background. The following observations will therefore look not only top-down, but also bottom-up. They will elaborate on:

- substantive criteria;
- administrative organisation; and
- legal protection.

I IPPC as a Substantive Requirement

1 *Basic Obligations – a New Legal Technique*

Regulatory law, as opposed to law instrumenting self-control, normally establishes substantive requirements of environmental protection which are to be observed by polluting actors.[1] These requirements have traditionally been addressed to administrators. The administrator, while exercising the power given to him/her to control the

---

* Professor of Public Law, Director Forschungsstelle für Europäisches Umweltrecht, Law Faculty, University of Bremen.

1 See G. Winter, 'Substantive Criteria of Environmental Protection', in: G. Winter (ed.), *European Environmental Law. A Comparative Perspective*, Dartmouth, Aldershot (1996), p. 37 et seq.

private party's activities by means of orders, authorizations, etc. must implement and respect those requirements. They at the same time empower and bind the executive, the underlying assumption being that the private party's freedom ends where those powers begin.

There is, however, a second possibility: the requirements may be framed as duties of care which are directly addressed to the individual. Besides obliging the individual they also serve as a reference for administrative control. The difference between the two models is not only formal. It seems that the second model fits better into a conception where the individual by that persons' own motivation (rather than only by governmental force) takes social responsibility upon him/herself. The duties of care give this motivation guidance even if the administrative agencies do not use their powers of implementation. One may object that if unenforced they are toothless, mere symbols without effect. But even symbols can have an impact, not to mention the possibility of connecting other juridical mechanisms to the observation of basic duties, such as tort liability or the obtaining of certain public benefits (subsidies etc.). Basic obligations can be especially useful for framing dynamic clauses such as the Best Available Techniques (BAT) standard which implies the duty to keep up with technological progress. They may also serve as reference criteria for the auditing of installations.

EC environmental law has traditionally followed the first model. It has used substantive standards as a means of directing the Member States' administrative action. By contrast, more modern German environmental laws such as the Federal Immission Control Act, the Act on a Circular Economy (the former Waste Act) and the Act on Genetic Engineering have adopted the second model.[2] For instance, Article 5 of the Federal Immission Control Act, paragraph 1 No. 1 prescribes that classified installations 'shall be established and operated in such a way that:

1. this does not involve harmful effects to the environment (...)
2. precaution is taken to prevent harmful effects to the environment, in particular by such emission control measures as are appropriate according to best available technology
3. residues are avoided (...)
4. any waste heat generated is utilized (...).'[3]

Article 3 of the IPPC Directive can be read as also introducing a catalogue of basic obligations to be established by the Member States. At least this is said in the title of the provision, and it can be inferred from sentence 2 which holds that compliance is sufficiently ensured if the conditions of permits are determined by the basic obligations. Although (over)cautiously limiting the legal extension this means, by implication, that the basic obligation possesses an excess moral value. This is not so much as a first step,

---

2 See for British examples the Environmental Protection Act (EPA) s. 34 (on waste-related duties of care) and s. 108 on duties of care in connection with gene technology.

3 For the complete version see G. Winter (ed.), *German Environmental Law. Basic Texts and Introduction* (1994).

but others, and more explicit ones, may follow, including the establishment of cross-references from the Auditing Regulation and a future Directive on Environmental Liability.

2 *Hazard Avoidance and Precaution*

Article 3(a) and (b) of the IPPC Directive lay down that no significant pollution may be caused, and that preventive measures must be taken. In drawing this distinction the article apparently resounds the distinction between the principles of hazard avoidance and precaution much cherished by German environmental law albeit not always to the satisfaction of those who call for clarity and simplicity. By adopting this distinction the Directive my be trapped by exactly the same doctrinal enigmas with which German lawyers have for a long time struggled.

One of the problems is how the dichotomy of hazard avoidance and precaution relates to the other one of quality objectives and emission limit values. Some have argued that precaution is always sufficiently observed if BAT are installed and other measures are taken, so that under these circumstances the denial of a permit is not a viable means of precaution anymore. The same and other authors have, in addition, reasoned that environmental quality objectives always reflect the principle of hazard avoidance.

These controversial issues may reappear on the Community level, because Article 10, which addresses these questions, does not answer them with sufficient clarity: do 'other measures' which shall not be prejudiced include the denial of a permit? Does the 'environmental quality standard' which may require stricter conditions than those achievable by the use of the BAT, include standards set at a precautionary level? If so, does this at least allow the requirement that the operator, short of being refused the permit, must take more stringent measures than those advised by BAT?

In the context of finding answers to these questions it may be of interest to note how they are discussed on the level of German law. I believe it is fair to say that the following is the prevailing opinion:

a) Before it can be said what measures are to be taken the situation must first be analysed and assessed. This *risk analysis and assessment* may reach three different results:

(1) The installation will, with high probability (which can be stated with sufficient certainty), cause significant harm to human health or the environment. It is then situated in the zone of hazard (*Gefahrenbereich*).
(2) There is only a low probability of causing harm, or else there is only a suspicion or tentative knowledge of it. The case is then classified as being situated in the zone of risk (*Risikobereich*). It may be noted that in the case of extraordinarily severe harm a low probability of occurrence suffices to classify the case as a hazard.
(3) There is no or an extremely low probability of causing harm, or in spite of relevant investigation, no knowledge whatsoever exists about possible adverse effects.
This related category is called the zone of residual risk (*Restrisikobereich*).

b) Different *consequences* are tied to these three zones:

(1) Hazard must be avoided (*Gefahrenvermeidung*). This means that safety measures must be devised and fixed by conditions in the licence in order to move the project downwards at least into the zone of risk. Should this not be possible, the installation cannot be authorized or, if already existing, can be closed down.
(2) Risk must be reduced, in particular through the application of the BAT (*Vorsorge*). Should, in spite of the application of BAT and other (e.g. operational) measures, the risk remain, the further course of action is controversial: some would regard this as a case of residual risk, others as still a risk situation which after further balancing (of the kind of affected environment, of the kind of installation, of the cost of removing it to another site etc.) allows the administration to finally deny the approval. The latter opinion strikes me to be the more flexible one. It may also be recommended for the interpretation of Article 10 of the IPPC Directive.
(3) Residual risk must be tolerated.[4]

c) How does this relate to the two modes of standardization, namely environmental quality objectives and emission limit values? The latter reflect the standard of BAT and thereby help to clarify what measures have to be taken in the risk zone. The former may either delineate the threshold between the hazard and risk zones, or draw a line within the risk zone including a line which separates this from the zone of residual risk. For instance, the Technical Instructions on Air Quality Control, an administrative guideline issued by the Federal Government, lays down a precautionary threshold of 60 μg SO_2 per m^3 of air which shall protect even spruce trees, a species known to be very sensitive to sulphur impact.[5] By contrast, the hazard threshold was fixed at 140 μg. The precautionary limit value if exceeded by an installation applying BAT can lead to the denial of an authorization or, in the case of an existing installation, to improvement orders.

It is submitted that Article 10 of the IPPC Directive could be interpreted in the same way. The provision does not specify the reference criteria for the cited environmental quality standards, but these standards must be seen as one of the 'necessary measures' the Member States shall take according to Article 3, and which do refer to both hazard avoidance *and* risk reduction. The reference made by Article 9 paragraph 4 to the 'local environmental conditions' cannot be regarded as obstructing the quality standards of any of the two kinds; such considerations only apply within the framework of the existing standards, or in cases where like standards have not been established.

4 As an exception in the area of nuclear energy law one speaks of risk minimizing measures to be taken even in this zone. But I regard this as an unnecessary further complication of the already very complicated scheme. In fact, such measures can well be placed in the risk zone. The intention behind the exception is to exclude third-party litigation. But this could be dealt with by other means.

5 No. 2.2.1.4. An English version of the guideline is reprinted in Winter, op. cit. (note 3), p. 199.

3 Best Available Technique

BAT as defined by Article 2 No. 11 allows for economic considerations of two kinds: One postulates 'economically viable conditions' for the 'relevant industrial sector', the other 'costs and advantages'. The first may be called affordability (for the sector). It means that a technique is not regarded as being 'available' if, for instance, a steel producing firm due to the worldwide competition pressure, just as its competitors, cannot afford the purification of its sewage. The second kind of economic consideration is a variant of the proportionality principle. It means that a technique is not viable if, for instance, its employment for removing the last bit of lead from waste water would cause extraordinary costs.

The definition of the related German term, i.e. *Stand der Technik* (state of the art),[6] although corresponding exactly with the first (technical) half of the BAT definition (most effective and advanced stage etc.), is inconsistent with the second (economic) half of the definition. Whilst the proportionality principle may be read into the German definition of *Stand der Technik* as an unwritten addendum (which on constitutional grounds permeates the whole legal order) affordability is clearly no requirement. It has, it is true, been a precondition for subsequent improvement orders but was abandoned because it was felt to prevent the administrative agencies from making use of their powers.[7] In any case it has never been an element of the definition of *Stand der Technik*.

The German way of defining BAT has constituted a cornerstone of the precautionary principle, which makes it understandable that the new EC formula has met severe criticism by German authors.[8] It is true that Article 130t of the EC Treaty allows Germany to preserve its stricter standard, but given the current international trend of 'racing to the bottom', German legislation may face pressure by the industry to yield to the European standard.

4 Integrated Pollution Control

The integration approach to pollution prevention and control is most clearly stated in consideration Nos. 6–9 and Articles 3, 9 of the Directive. Whilst the Latin origin of the word simply connotates improvement, embetterment, integration deriving from *integrare, integer*, the current understanding means the seeing and managing of something (which procedure will, hopefully, indeed improve the situation). What the Directive likes to see

6 See Art. 3 para. 6 Federal Immission Control Act.

7 It was replaced by the precondition that the improvement order must not cause excessive cost the operator bearing the burden of proof for the non-proportionality. See Art. 17 Federal Immission Control Act.

8 See D. Sellner, J. Schnutenhaus, 'Die geplante EG-Richtlinie zu "Integrated Pollution Prevention and Control (IPC)"', *Neue Verwaltungsrechtszeitschrift* (NVwZ) (1993), p. 828 et seq.; I. Appel, 'Emissionsbegrenzung und Umweltqualität', *Deutsches Verwaltungsblatt* (DVBl.) (1995), p. 399 et seq.

to be managed in an integrated manner is the various kinds of discharges from an installation, including emissions (in the widest sense) to the air, to the ear, to the water, to the soil, to the waste heap. No 'problem shifting' shall occur. In addition, the energy input must be used efficiently. All this is designed to protect the 'environment as a whole'.

If one regards the Environmental Impact Assessment (EIA) Directive as creating a mere procedural device to assess the effects on the environment as a whole one might see the IPPC Directive as its substantial and instrumental complement.

Looking at the related German law some steps towards an integrated concept can be found although others will still have to be taken in the process of transposing the Directive.

A major step was taken by the Federal Immission Control Act which regulates classified installations. In terms of the scope of the media to be considered it is sufficiently broad: not only the air but also water, soil, the atmosphere, animals and plants are included. But the Act covers only a limited number of causative paths, namely 'imponderable emissions' like exhaust fumes and noise. The Act also lays basic duties relating to waste, waste water and waste heat.[9] More detailed duties as to waste and waste water, e.g. regarding the discharge of waste water and the extraction of process water are, however, a concern for special laws on solid waste and on water management and protection. The same is true for duties regarding physical encroachment on nature and landscape, and for regulations under the land-use planning and construction laws.

In spite of this diversity as regards the legal sources of the relevant duties, many of these duties have to be applied by the authority responsible for the immission control licence which thereby takes a leading role. To the extent that the other laws require a special licence (e.g. the zoning and construction laws) this licence is even comprised by the leading licence provided under the Federal Immission Control Act, a doctrinal construction which is called concentration effect. There are, however, important exceptions to this concentration. Those are the licences regulated by the water and mining laws.[10]

With regard to the integration clause in the narrower sense there is in the German laws no clear stipulation of the principle of best practicable environmental option established by British law.[11] However, Article 2 of the German Environmental Assessment Act comes close to this by requiring that:

9 See above section I 1 (*supra*). For details on the integration elements already to be found in the Federal Immission Control Act see M. Rebentisch, 'Die immissionsschutzrechtliche Genehmigung-Ein Instrument integrierten Umweltschutzes?', in *NVwZ* (1995), p. 949 et seq.

10 See Art. 13 of the Act. By contrast, under British law not only the water use permit but also the planning permission are excluded from the integrated licence. See P. Castle, H. Harrison, *Integrated Pollution Control* (1996), p. 162. See also R. Macrory in his contribution to this volume.

11 Section 7(7)(b) EPA.

> if the approval of the project is decided upon by several procedures, the individual assessment carried out in these procedures are compiled to provide an overall assessment (*Gesamtbewertung*) of all environmental impacts, including interactions (*Wechselwirkungen*).

This requirement is understood to forbid problem shifting.[12] In addition it is taken to mean that should an installation meet all *individual* environmental quality standards it could nevertheless be considered to be harmful if the *overall effect* is considerable.[13] Unfortunately, the Environmental Assessment Act being by most scholars interpreted to be only a procedural instrument, the said clause is not given much substantial bearing. A version with substantive effect is presently being considered by the Independent Expert Commission for an Environmental Law Code. It reads as follows:

> The basic duties related to the project and the requirements for the admissibility of encroachments upon nature and landscape are to be observed in such a manner that considering all of the causative chains and the mutual interaction of the environmental elements such measures are taken which carry the least burden for the environment as a whole.[14]

II Administrative Organization

1 *Authorization of New Projects*

Integration of the substantive evaluation of a project must be reflected in administrative organization if it shall become effective. Therefore, Article 7 of the Directive demands the following: 'to ensure that the procedure and conditions for the grant of the permit are fully coordinated where more than one competent authority is involved'.

This is a rather modest request. It allows that for one project several permits are required to be issued by different agencies through different procedures. Normally, the coordination mandated by the cited article would require a lead agency to be appointed.

---

12 An example for problem shifting which has frequently been cited is flue gas desulphurisation in coal-fired power stations. The removal of SO_2 produces high quantities of plaster. To the extent that this cannot be reused it must be deposited, which leads to chloride pollution and soil impairment.

13 See the English version of a communication of the *Bundesumweltministerium* of November 1994 entitled 'Problems in the Implementation and Enforcement of Directives of the European Union in the Field of the Permitting of Industrial Installations', at p. 20. See also W. Erbguth, A. Schink, *Gesetz über die Umweltverträglichkeitsprüfung. Kommentar* (1992), § 2, Nos. 30–33.

14 Ministry for the Environment, Nature Conservation and Nucelar Safety of the FRG (ed.), Environmental Code, Draft, Berlin 1998 (ISBN 3–428–09227–9). The cited provision is to be found in Art. 83 para 2.

This modest variant of organisational integration has, in Germany, been adopted for the general EIA procedure. One lead agency which is to be designated by *Land* regulation at least determines the scope and produces the summarized description of the project effects but may be given more competences, in particular as to the evaluation of effects.[15] In any case, the different agencies remain competent for conducting their respective licensing procedures, the taking into account of the Environmental Impact Statement (EIS), and the granting of the permit.

There is a stronger variant of administrative integration in German law which may be worth looking at. It is characterized by what is called concentration effect *(Konzentrationswirkung)*. Concentration effect means that the several permits which various laws may require for one project are embodied in only one permit. This permit is issued by one lead agency. The other agencies' role is confined to them being heard in the decision-making procedure. In some cases their consent must be obtained by the lead agency.[16] The lead agency when taking its decision has to comply with the substantive criteria established by the various other laws.

One example of such concentration effect (or rather of a partial concentration effect) is the already cited permit regulated by the Federal Immission Control Act.[17]

An even stronger variant of administrative integration is presently being discussed by the Independent Expert Commission for an Environmental Law Code. In addition to framing an integrated set of substantive duties of care including the above-cited Best Practical Environmental Option (BPEO) principle the Commission has, as an organisational complement, developed the integrated permit. This is constructed not as an umbrella permit 'concentrating' several individual permits but as a one for all decision.[18] The logical consequence for the organisational structure is that the specialized agencies for air, water, soil, etc. are replaced by one agency competent for the environment as a whole. This consequence is, however, not unavoidable. Also consistent with the concept of the integrated licence is a solution where the specialized agencies are retained and made to be heard in the licensing procedure which is guided by the agency responsible for the integrated licence. The Commission leaves this to the *Länder* to decide.

In my mind, the second model is preferable.[19] The reason is that a specialized agency develops a longitudinal perspective on its medium which allows it to assess how much additional burden the medium may be able to tolerate. It will also identify itself

---

15 See Art. 14 of the Environmental Impact Assessment Act, reprinted in the collection cited in note 1.

16 This applies to the water authority in cases of plan approval for infrastructure projects such as highways. See Art. 14 of the Federal Water Management Act.

17 See above s. I 4.

18 Art. 99 of the Draft Code (see note 14).

19 See for a more elaborate discussion of this view P. Knoepfel, 'How to Organize Environmental Administration', in: *European Environmental Law*, op. cit. (note 1), p. 157 et seq. as opposed to G. Lübbe-Wolff, *Modernisierung des Umweltordnungsrechts* (1996), p. 59, who argues in favour of an integrated approach.

with its entrusted part of the environment. By contrast, the integrated agency will tend to develop cross-cutting perspectives. Not identifying itself with special media it will be more flexible to sacrifice one to the benefit of another depending on what the project requires. This will be all the more so if the agency is integrated even in such a way that it also has responsibility regarding the economic side of the project, as, for instance, the need for a new highway or a new power plant. Experience shows that the 'softer' realms like in particular nature and water protection, are better represented in the decision-making if they are handled by separate agencies, especially if these enjoy a veto power. If composed as a division of an integrated super-agency they are more easily outplayed by more powerful concerns.

2 *Supervision*

Environmental lawyers tend to concentrate on authorization problems and underestimate questions of supervision and improving existing installations. This is contrasted by the fact that effective supervision and improvement is probably much more efficient in terms of environmental protection gains per working unit of administrative personnel.

The Directive instructs the Member States in both respects. According to Articles 13 and 14 they shall ensure that the permit conditions are updated and complied with.

As a technical aside it may be noted that the provisions could have been framed in such a way that the basic duties established by Article 3 are taken as a point of reference for the reconsideration and updating of the permit conditions.

What is more important in the present context is that the Directive does not say anything about the impact of the integrated approach on the control of existing installations. This is not to be criticized because organisation is a matter for Member State legislation. Nevertheless, the question arises, and has to be answered by the Member States, of what agency should be responsible for supervision and improvement measures if the initial permit was granted by a lead agency. Should the specialized agencies regain full powers? Certainly not insofar as this would lead to the dismemberment of the coordinated, concentrated or even integrated licence. If, for instance, the agency responsible for water protection finds that the polluting installation should be brought up to the new BAT level it is doubtful whether the agency should be able to command this by a subsequent improvement order because the order might contradict other permit conditions and impair the BPEO balance embedded therein. Therefore, in a system of coordination of different permits the water agency would have to consult with the other agencies in order to ensure that the integrated solution is not jeopardized. In a system of both concentration and integration effect the water agency would not be competent to issue subsequent orders but would rather have to ask the lead agency to act.

The situation can be seen differently in cases of enforcement. Should the water agency discover non-compliance with the conditions of the permit it may well retain its powers to issue orders appropriate to force the offender to desist from committing the offence, including the temporally closure of the facility. If this proves unsuccessful and the revocation of the permit must be envisaged, the competence must logically revert to the lead agency.

In federal systems the situation may be complicated by the fact that both Federal and *Land* agencies may have competences in a specific field. But, in principle, the division of labour between the lead agency and specialized supervisory agencies applies accordingly.

3 *Participation of the Public*

Integration can also be understood to refer to the public, i.e. to incorporate into the process the individuals interested in and affected by the project, including persons living in neighbouring Member States. This may be the reason why the Directive takes up questions of public participation although these might have been left to the revision of the EIA Directive the thrust of which is basically on procedures.

As to the participation of the public living on German territory the national law can be considered to be already in compliance with the requirements set up by Article 15 of the Directive, provided the list of classified installations is adapted to the Annex I list of the Directive. There is only a small point to add. It concerns the relationship between public participation under the IPPC Directive and freedom of access to environmental information under Directive 90/313/EEC. Britain, for instance, denies additional access rights pending formal public participation procedures.[20] Going further, Germany denies such rights even pending non-formal administrative procedures.[21] The effect is, in the case of public participation procedures, that third parties are confined to consulting the documents put on display. They cannot request further information, e.g. information received by the responsible agency after the termination of the period of display.

This arrangement is said to be grounded on Article 3 paragraph 2 3rd indent of Directive 90/313/EEC where it is said that matters which are the 'object of preliminary procedures' may be exempted from the free access right.

Whilst such a broad interpretation of 'preliminary procedures' is in itself doubtful (preliminary meaning, in accordance with the word's origin in the French Act No. 78–753, Art. 6, preliminary to court proceedings),[22] Article 15 of the IPPC Directive gives an additional hint in the same direction for what else the qualification 'without prejudice to Council Directive 90/313/EEC' could mean, but that even in the case of public participation procedures the right of access to information prevails.

20 See Environmental Information Regulations 1992, Stat. Instr. 1992 No. 3240.

21 Art. 7 para. 1 No. 2 of the Environmental Information Act. In a recent judgment the ECJ has held that this is to be in violation of Art. 3 para 2 3rd indent of Directive 90/313/EEC. See ECJ judgment of 17 June 1998, case C–321/96.

22 According to Art. 6 of the Act the administration may refuse access, if this could 'porter atteinte au déroulement des procédures engagées devant les juridictions ou d'opérations préliminaires à de telles procédures, ...'.

Referring to the participation of the foreign public it may be noted that in contrast to the EIA Directive[23] the IPPC Directive extends the duty to inform not only to the other Member State but also to the public of this state.

Different methods of discharging this duty can be imagined. One, which was chosen by Article 17 paragraph 2 is to make the neighbouring state responsible for informing its public, the emitter state being obliged to allow the foreign public to comment before the permit is issued. Another solution could be that the public of the receiver state must be directly informed by the emitter state. This, it is true, would imply that the emitter state exerts governmental functions on the territory of the other state. The territoriality principle is, however, in manifold ways relativized by various mechanisms within the EC, including, for instance, the principle of origin which requires Member States to respect licences for products and services issued by other Member States. Besides, the official publication of information about a project could be replaced by private advertisement in the media.

III Legal Protection

The IPPC Directive does not touch upon questions of the legal protection of those affected by administrative practice under national law emanating from the Directive. This corresponds to normal Community practice: Legal protection is by and large a matter of national legislation. However, one crucial point shall be raised which must be expected sooner or later to arise in the German context.

Emission limit values are of crucial importance for the integrated pollution prevention conception for by them best environmental practices can be put into concrete standards. They may be set as conditions to the individual permits[24] or as general values. General values set at the national level must respect Community values which may be established by Council Directive in accordance with Article 18 or which, in their absence, already exist within or on the basis of the various Directives listed in Annex II of the IPPC Directive.

Let us assume that a limit value was correctly transposed into German law but not applied by the competent authority in a permit granted for the construction of a factory. Can a neighbour of the plant seek legal protection against the permit for an excess of the emission limit?

According to the German principles of standing an individual has only a right to sue the authorities to properly implement the principle of hazard avoidance. Such a right is not conceded with regard to the principle of precaution. The reason for this is to be found in a construction characteristic of German public law, i.e. the theory that the relevant law must aim at the protection of individual persons rather than solely

23 In the meantime this difference was eliminated by an amendment to the EIA Directive, see Directive 97/11/EC, OJ L 73/5, Art. 1 No. 9.

24 See Art. 9 para 3 of the Directive.

at the benefit of the general public. This effect must be the intention of the law, mere factual impact is not sufficient. This theory is called the theory of protective norms (*Schutznormtheorie*).

According to the prevalent opinion the principle of prevention is meant not to protect individuals but rather the public health as a collective good. The principle also allows reserving high protection zones for nature preservation and possible future development. These intentions are seen to be interests of general welfare.[25]

In terms of our case the result of this reasoning is that the neighbour cannot invoke the emission limit value as such. He/she can do this only in cases where the excess of the emission limit value at the same time constitutes a violation of the principle of hazard avoidance. This proof will in practice be alleviated if there are environmental quality objectives which specify the hazard level.

The question which arises in the Community context is, whether this result is compatible with the Community emission standards. Although the shape of the legal protection including the construction of standing is a national matter the *effet utile* requires that a minimum standard be met.[26] To the extent that a Community legal act establishes individual rights these rights must in principle be made pursuable by the national courts (which is not to say that e.g. deadlines for a claim, the right to *certiorari*, the existence of summary proceedings, the fee structure etc. must be identical).

Under what conditions a right is established is of course open to discussion. In contrast to rules of standing in the context of Article 173 paragraph 4 EC Treaty, where the factual effect of the Community measure is relevant, in the context of Article 169 proceedings the ECJ applies a version of the *Schutznormtheorie*, i.e. the Court poses the question what the intention of the legal act is, and whether the act aims at protecting the individual citizen. Looking at the case law it is probably fair to say that the Court in assuming an individual protective intention is more generous than German courts and possibly also other Member States courts.[27] For instance, the Court has accepted such intentions in the cases of environmental quality objectives for lead[28] and of quality criteria of drinking water,[29] although the protected individuals are obviously very numerous in these cases.

For emission values no jurisprudence of the European Courts appears to exist. The answer will probably be found in an analogy to the judgments concerning quality objectives: the individual legal act must be scrutinized with respect to formulations which declare its protective aim. For instance, Directive 85/210/EEC which sets emission values for lead in gasoline declares that this is done in the interest of public

25 Bundesverwaltungsgericht, Judgment of 18 May 1982, BVerwGE 65, 313 et seq., at 320. For a critical discussion see E. Rehbinder, in: *Festschrift für H. Sendler* (1991).

26 See e.g. cases C–6 and 9/90 (Francovich), Rep. 1991, I–5357, at 5415.

27 J. Jans, *European Environmental Law* (1995), p. 124; I. Pernice, 'Kriterien der normativen Umsetzung von Umweltrichtlinien der EG im Lichte der Rechtsprechung des EuGH', in *Europarecht* (1994), p. 325 et seq., at 339.

28 Case C–59/89 (*Commission* v. *FRG*), Rep. 1991, I–2607 et seq., at 2631.

29 Case C–58/89 (*Commission* v. *FRG*), Rep. 1991, I–4983 et seq., at 5023.

health, in particular in regions with high traffic density. This amounts to a subjective right at least for those who actually live in such regions.

By contrast, the IPPC Directive does not specifically mention the health of individuals. It frequently refers to the environment as its protected good and assures that a high level of protection is intended. However, to infer therefrom that individuals are not within the protective scope of the Directive sounds a bit positivistic. After all, in the first consideration it is said that the integrated pollution prevention shall serve the goals laid down in Article 130r EC Treaty, part of which is human health. Given the more generous version of the Court's *Schutznormtheorie* this would probably suffice for deriving subjective rights from the Directive. This of course does not mean that everybody can claim to be affected in his or her right. For this the possibility of being harmed must be substantiated. But this is a question for the national legal system. The same is true with respect to the density of court review. In German law court review concerning the prevention principle would avoid replacing the administration's assessment by its own discretion.

It may be added that the instigation of court review with regard to the prevention principle might rather be a case for the association action. In Germany the association action in the environmental field is reserved to the protection of nature and landscape for these goods are in general not allocated to individuals. On the basis of a narrower conception of the *Schutznormtheorie* one could argue that the good protected by the prevention principle is indeed a general one, i.e. public health, air, water and soil as such, land use planning interests etc. We would then have another collective good before us for which another form of association action may be an appropriate remedy in cases of non-respect of the prevention principle.

IV Summary

1 *On Substantive Criteria*

The introduction of basic operator duties (Art. 2a) is uncommon in EC law but rather familiar in German law where, however, they lack doctrinal clarity.

The relationship between the emission and environmental quality perspective (Arts. 2a, 9), while improved by the Common Position, is still unclear in cases where additional measures should be required but are not viable. Neither is German law clear in this respect.

The scope of impacts to be considered under German law is broader than assumed by those who consider German law to have no IPC tradition. The scope is, however, narrower than required by the Draft Directive.

By referring to economic viability and to proportionality of costs and advantages the definition of BAT (Art. 2 No. 10) is more 'economy-minded' than the corresponding German term *Stand der Technik*. With respect to Article 130t EC Treaty Germany is, however, not forced to adopt the less stringent standard (although this will probably be done in practice).

As to the BPEO principle Article 8 may be read in order to establish that German law does not accept this but for the area of risks. No trading of damage to one medium for the sake of protecting another medium is allowed.

2 *On Administrative Organisation*

Three degrees of organisational integration may be distinguished from a German perspective:
- coordination of procedures; several permits;
- coordination of procedures; one permit with 'concentration effect';
- unified procedure; one permit with 'integration effect'.

While the Draft Directive has adopted the first variant, German law mixes the first and the second. The Environmental Law Code as drafted by a ministerial commission, will propose the third variant. A necessary problem is the potential loss in longitudinal perspective on the individual media.

The Draft Directive does not tackle the problem of which agency is responsible for what enforcement and improvement orders in cases where one permit integrates others. Experiences with German schemes of concentrated permits may help in solving such problems.

3 *On the Participation of the Public*

German law rather corresponds with the requirements contained in the Draft Directive as to formal procedures and access to information. Incidentally, an interpretive clue may be found with respect to the reading of 'preliminary procedures' in Directive 90/313. The rules concerning transboundary participation will need some improvement on the part of procedural law, in particular with regard to cross-border notice, comment and hearing procedures.

4 *On Legal Protection*

Whilst legal protection is left to Member State regulation the Draft Directive might nevertheless indirectly challenge traditional national concepts, such as the following German ones: Can emission limit values or only environmental quality objectives be invoked by third parties? How does the BPEO principle affect the narrow German conception of standing? Ought a 'precautionary' association action be introduced?

V Conclusion

European conditioning of national laws often leads to innovation, sometimes even to systemic change, but in some other cases to mere molestation or even regression. Against this background the IPPC Directive, if taken at face value, must be termed innovative. Amazingly enough it reinforces administrative control in times where the overall trend points in the direction of self-regulation. While in the author's mind this is more of an advantage, some scepticism remains: The BAT definition and the BPEO principle as well as integrated organisational schemes established on the Member States level are not immune to being practised in favour of more latitude for polluting activities.

Integrated Pollution Prevention and Control: the Dutch Experience

Th.G. Drupsteen[*]

I Introduction

In the flyer promoting the congress, this chapter was given the somewhat ambiguous title of 'The Dutch experience'. I do not know whether you have ever had a Dutch experience, but it does not sound very positive to me! A Dutch experience can easily be associated with Dutch courage, a Dutch treat or Dutch comfort. So I hope you will understand why I feel a little uncomfortable about beginning a chapter with this title.

The title was chosen by the organizers because the intention was that two countries should be compared, one without a tradition of Integrated Pollution Control and one, supposedly, with such a tradition. To make things even more complicated, the organizers thought that it would be interesting to choose the Federal Republic of Germany and the Kingdom of the Netherlands as the two countries. Now you all know that the relationship between Germany and the Netherlands is a very intense one, but also a delicate one. I am therefore loath to ruffle any feathers by suggesting that we in the Netherlands have something that the Germans do not have, even something as minor as a tradition of IPC, when in fact we do not.

So I am rather reluctant to start talking to you about the Dutch experience, suggesting that we in our country do have a long-standing and rich tradition of IPC, when that is simply not true. The Dutch Environmental Management Act, which imposes integrated pollution control by making industrial installations subject to integrated licensing requirements, only came into force a few years ago on 1 March 1993. Little systematically compiled information has, as yet, been made available about the way this licensing system works in practice, about its advantages and disadvantages. Besides, as you will see, our 'integrated' licensing system is really only semi-integrated, because some forms of pollution are still covered by other legislation, separate from the

* Professor, University of Leiden.

Environmental Management Act, and one form of pollution is subject to separate licensing arrangements, which are simply coordinated with the more general Environmental Management Act system.

What I intend to do is to provide a short historical survey of the development of integrated pollution control in the Netherlands; then I will deal with integrated environmental licensing under the Environmental Management Act in more detail, trying to provide some idea of how the system works in practice. At the end of this Chapter I will come to the question of whether the IPPC Directive, which is still in draft form, will call for the adoption of this exciting Dutch approach.

II Historical Development of IPC

One might reasonably suggest that in the Netherlands the control of pollution from industrial installations has been organized on an integrated basis for a long time. From 1875, industrial pollution was governed by the Nuisance Act, preceded by Napoleon's Imperial Decree in 1810 and by Royal Decree in 1824. The Nuisance Act covered all kinds of nuisance. It has been argued that the function of the Act was more symbolic than practical. Investigations during the 1970s revealed that only about 25 per cent of the industrial installations which were supposed to be licensed under the Act actually had fairly up to date licences. The Act's licensing system dit not really work in practice and enforcement was almost non-existent.

Under the Nuisance Act, it was the municipal authorities that were responsible for licensing. In the second half of the 1960s, public concern focused on new forms of environmental pollution and the government came to the conclusion that the Nuisance Act was not an adequate basis for an effective modern environmental policy. The Act covered only pollution from industrial installations, while the municipal authorities, being concerned with fairly small areas, were not in a position to deal with the more important cross-boundary forms of pollution. The difficult decision facing Dutch policy-makers at that time was whether to set up a completely new integrated environmental protection act covering all the different forms of pollution or to develop a number of separate acts, one for each form of pollution. The Dutch Government deliberately chose the second alternative. Effective control of the most serious forms of pollution was considered to be so urgent that there was not enough time to develop a general environmental act covering a wider range of pollutants.

It was expected that separate environmental acts for different forms of pollution could be brought into effect much sooner. From that time, about 1971, an integrated approach to environmental pollution was deliberately abandoned in our country, and a number of sector-oriented environmental acts came into force. Each of them had its own licensing criteria, with most but not all placing the provincial authorities in charge of licensing; the licensing procedures showed slight but sometimes important differences and, because the legislation was sector-oriented, central and local government administration was also organized on a sector by sector basis.

It cannot be said that the choice of a sector-oriented approach was completely unsuccessful. On the contrary, real progress was made in the control of surface water pollution, air pollution and traffic noise. Nevertheless, the disadvantages of a system of sector-oriented environmental legislation soon became clear. Three important drawbacks spring to mind. First, there were the slight, mostly technical differences in regulation, that caused confusion for administrative bodies, licence applicants and other interested parties. Second, as the number of sector-oriented acts increased, so the number of cases increased in which there was serious doubt as to whether a given activity was governed by one act or another, or by both. Uncertainty regarding the applicability of various acts was particularly problematical when different administrative bodies were responsible for licensing under the acts in question. Third, a system of sector-oriented environmental acts was found to hamper an integrated approach to the different forms of pollution from one installation. Every form was examined according to the criteria of the relevant act and there was no opportunity to review the accumulated effects of all forms of pollution or even to balance an increase in one form of pollution against a decrease in another.

Although some of these disadvantages were already being mentioned in the mid 1970s, it was not until 1980 that the technical differences were addressed, and we had to wait until 1993 for a solution to the other two problems. In 1980, the Environmental Protection Act, General Provisions came into force. This Act was only a small step towards a system of integrated pollution control; it may be regarded as more of a legal and technical harmonization of environmental legislation than material integration. The Act prescribed a uniform licensing procedure for nine different sector-oriented acts. It also made provision for coordinating the licensing of a given installation, although this provision was not used very frequently. The Environmental Protection Act, General Provisions was later extended to cover matters such as environmental impact assessment and financial provisions. However, the problem of deciding what legislation was applicable in complicated cases and the absence of any way of taking the accumulated or integrated effects of pollution into account remained unsolved until 1993.

III Integrated Environmental Licensing

The great step forward was made in 1993 when a new system of integrated environmental licensing came into force, together with a system of integrated environmental policy planning, regulation of environmental quality standards and regulation of enforcement by administrative instruments. The existing licensing procedure was simplified and speeded up. The new arrangements were used as the model for a more general procedure for taking administrative decisions regarding applications, which came into force on 1 January 1994 as a part of the General Administrative Law Act.

The provisions which form the basis of the Netherlands' integrated environmental licensing system are contained in sections 8.1 and 8.10 Environmental Management Act. Section 8.1 makes it illegal to establish or alter an installation, to alter the operation of an installation, or to operate an installation without a licence. The categories of installation subject to compulsory licensing are specified in the Establishments and Licences Decree. Under this decree, the municipal authorities are in principle responsible for licensing, but, in most categories, if the size, production capacity or number of the installation(s) exceeds a set threshold, the matter has to be referred to the provincial authorities.

The effect of this system is to make the municipal authorities responsible for licensing smaller installations of the kinds previously covered by the Nuisance Act, and the provincial authorities for the larger ones, which in the past were governed by the Air Pollution Act, the Noise Abatement Act and the Waste Disposal Act.

Section 8.10 states that a licence can only be refused in the interests of environmental protection. Under section 11, which leads on from section 10, licences can be issued subject to restrictions or conditions necessary for the protection of the environment. The Environmental Management Act itself defines in section 1.1 what is meant by the protection of the environment. The phrase covers improvement to the environment, effective disposal of waste, economical use of energy and materials and protection of the environment against the detrimental effects of transporting people and goods to and from the installation. So an environmental licence issued in accordance with the Environmental Management Act not only deals with the prevention or restriction of traditional types of pollution, but also with measures to improve the condition of the environment, with the disposal of waste from the installation, the conservation of energy and materials and with the traffic to and from the installation.

The municipal and provincial authorities do have certain discretionary powers regarding the imposition of restrictions or conditions. The Environmental Management Act is, however, designed to restrict these discretionary powers in certain respects. First, the Act states that if the environmental impact of the installation cannot be negated by the imposition of conditions, licensing should be subject to conditions designed to reduce the impact as far as possible, unless such conditions would be unreasonable. Thus, the so-called Alara (as low as reasonably achievable) principle is embodied in the Environmental Management Act. The essence of the Alara principle, originally applied in the field of radiological protection, is that exposure to radiation, if unavoidable, should be as low as reasonably possible. Until now, the practical implications of applying the Alara principle to other – perhaps less harmful – forms of pollution have not been entirely clear. To what extent should the costs of prevention or reduction be taken into account, for instance, and does the principle differ in its effect from an insistence on the use of the Best Available Techniques or even the Best Practicable Techniques?

Second, the Environmental Management Act gives a set of criteria, which variously have to be applied, taken into account or considered by the licensing authority when assessing a licence application. The authority's decision is based most loosely on those criteria which only have to be considered and most firmly on those which have to be applied. The criteria which have to be *considered* by the licensing authority are:

- The present condition of the environment in respects which are potentially subject to the influence of the installation.

- The potential impact of the installation upon the environment.
- The way the installation and the area in which it is located are likely to develop in the future, insofar as such developments may reasonably be predicted and are relevant to the protection of the environment.
- The comments and recommendations received in response to the draft licence.
- The scope for protecting the environment by negating or restricting the impact of the installation.

The criteria which have to be *taken into account* are:
- The content of the applicable environmental policy plan.
- The quality targets for those environmental factors which the installation might affect.

Finally, the criteria which have to be *applied* are:
- The existing quality standards for those environmental factors which the installation might affect.
- The general rules applicable to the category of installation in question, contained in either governmental decrees or provincial regulations.
- Relevant instructions issued by the minister responsible for Environmental Management.

Although the Act aims to restrict the discretionary powers of the licensing authorities in this way, one wonders whether such a detailed set of criteria will ever be effective. The idea was that by specifying assessment criteria in this way, a court could determine whether the licensing authority had acted correctly in the event of an appeal. However, a number of the criteria are so vague and the link between them and the decision so weak, that it is not always possible for a court to determine whether a decision has been properly arrived at.

In practice, other regulations, by-laws and policy rules perhaps restrict the discretionary powers of licensing authorities in more than a number of the above-mentioned criteria. I have indicated that general rules applicable to the installation, contained in either governmental decrees or provincial regulations had to be applied. I should elaborate on this a little. As well as requiring integrated environmental licensing, the Environmental Management Act broadened the system of general rules laid down in governmental decrees passed under the Nuisance Act, which applied to certain categories of installation whose environmental impact is generally small. It was considered more efficient to regulate these installations by a set of general rules than to require individual licensing of every installation in the relevant categories.

The categories concerned include office buildings, bakeries, butcher's shops and dairy farms. As part of a general policy of deregulation, the Dutch Government intends to increase the number of categories of installation regulated by general rules and to reduce the number of installations for which individual licensing is required as far as possible.

As far as I can see, this move towards wider application of general rules for certain categories of installation does not conflict with the provisions of the IPPC Directive requiring individual licensing of those installations mentioned in Annex I, which in general are the larger ones. The use of general rules may involve complete or partial removal of the licensing requirement. In the latter case, installations are regulated by a combination of general rules and licensing requirements. This system was in fact employed by some of the old sector-oriented legislation. The Air Pollution Act, for instance, required individual licensing but also required gas and oil installations to meet general product standards and large furnaces to meet general emission standards.

Finally, in almost all cases, licence applications are assessed on the basis of unofficial policy rules, instructions, recommendations and guidelines. In the event of an appeal, Dutch administrative courts tend to consider such guidelines to be more or less equal to official regulations or by-laws.

To conclude my short description of the integrated environmental licensing system, I should mention some of the system's limitations. Most important, permission to discharge into surface waters, as required under the Pollution of Surface Waters Act, is not covered by the integrated environmental licence. The Environmental Management Act and the Pollution of Surface Waters Act only contain provisions for the coordination of the two licensing systems. The background to the failure to integrate in this area is too complex to describe in detail here. The official line is that if the surface waters discharge licensing system were brought within the scope of the Environmental Management Act, the resulting integration of environmental policy would be at the expense of water policy integration. This has to do with the fact that discharge licensing is normally the responsibility of the water boards. The traditional function of the water boards was to provide protection against high sea and river water levels and to control inland water levels; in other words, water *quantity* management. During the seventies, the water boards additionally acquired responsibility for water *quality* management. Over the last decade, integrated water management concepts and methods have been developed, in much the same way as the integrated environmental management system has been developed. The water boards in particular were not very keen to see water management integration sacrificed to achieve greater integration in environmental management. Privately, many people believe that the water boards were simply unwilling to give up any of their power.

The second limitation is that none of the normal criteria on which applications have to be assessed apply to establishments producing one particular form of environmental pollution, namely acidifying ammonia from the manure of cattle. Where such pollution is concerned, special criteria apply, relating to the number of cattle in relation to the area of the farm, while the technical conditions applicable to the farm are set by a special act, namely the Interim Act on Ammonia and Cattle Farming.

Third, the process of integrating environmental legalisation in the Netherlands is only half complete. Important environmental acts, such as the Soil Protection Act, the Hazardous Substances Act and the Pesticides Act have yet to be integrated into the Environmental Management Act. Others, such as the Noise Abatement Act and the Air Pollution Act are only partly integrated. Most of these Acts either do not impose licensing requirements or, where they do, they have been integrated with those of the Environmental Management Act. However, the Pesticides Act has a system for admitting new pesticides, while the Hazardous Substances Act requires prior notice to be given before one is entitled to introduce a new substance to the market.

IV Preliminary Evaluation Results

The new environmental licensing system was expected to improve the regulation of industrial installations. However, we do not yet know whether it really does so. After all, the system only came into force on 1 March 1993, and sufficient time has not elapsed for serious evaluation. So I can only give you some initial impressions of how the system has been working in practice.

A certain amount of feedback has been provided by a comparatively small survey of about 70 medium-sized companies, commissioned by the Environmental Office of the Employers' Unions. The results of the survey relate mainly to the procedural aspects of licensing and the burden of information provision placed upon the applicant. Although the application procedure is supposed to take six months (a month less than the procedure operated under the Environmental Protection Act, General Provisions), in 55 per cent of cases no decision was reached within this legally prescribed term. Indeed, on average it took 9.3 months to obtain a licence. Before drawing conclusions from this data, it should be recalled that the investigation concerned medium-sized and large installations, for which the procedure is necessarily more complicated. On the other hand, we also know that the formal licensing procedure is usually preceded by an informal one, which itself usually takes seven or eight months.

The investigation showed that there is a strong correlation between the length of the procedure and the submission of objections to the draft licence. If objections are raised, it is virtually impossible to complete the licensing procedure within the legally prescribed term.

Some of the survey findings give an initial indication of the real scope of the integrated approach. Although respondents complained about the amount of data required for an application, only 66 per cent of the companies said they had definitely submitted data on energy use, while 7 per cent had not and 27 per cent did not know. Some 72 per cent of applicants provided data on traffic (23 per cent did not) and nearly all (93 per cent) submitted data on waste.

It is interesting to compare these figures with data on the subjects covered by licence conditions. While 66 per cent of the applications contained data on energy use, only 41 per cent of the licences were issued subject to conditions regarding energy consumption. The corresponding figures for waste are 93 per cent and 91 per cent, and for traffic 72 per cent and 50 per cent; 55 per cent of licences were issued subject to conditions regarding the use of materials. It is quite difficult to interpret these figures. The fact that only 50 per cent of the licences were subject to traffic conditions, while information about traffic was given in 72 per cent of the applications, does not automatically mean that there is something wrong. In the 'other' cases, it might have been apparent from the information provided that conditions regarding traffic flows were pointless because of the low traffic densities. The only general conclusion one may draw is that the full scope of the Environmental Management Act is perhaps not yet being utilized by the licensing authorities.

The investigation also provided information about the cost of applying for a licence; legal costs typically amounted to Dfl. 28,000, while the average cost of preparation was Dfl. 100,000. In 53 per cent of cases, coordination with the discharge licensing requirements of the Pollution of Surface Waters Act was needed. In general, this did not cause serious delay. Finally, in about 20 per cent of cases, an appeal was lodged either by a third party (31 per cent of appeals were submitted by third parties), by the applicant (35 per cent), by an environmental group (15 per cent) or by another party.

The introduction of the Environmental Management Act was carefully prepared by the ministry responsible for environmental management. Since March 1993, the ministry has also been trying to monitor implementation of the Act as closely as possible. It therefore started a project called VIEW. The 'English' title might suggest that the project has a kind of international flavour, but VIEW is just a Dutch acronym, which stands for '*Verzamelen van Informatie en Expertise Wet Milieubeheer*', in English: 'Collection of Information and Skills for the Environmental Management Act', which would give rise to a completely different acronym. As it happens, the first official results from the VIEW Project – which is not intended to collect quantitative data – have recently been published. The first VIEW report describes experiences with and opinions about the implementation and effects of the Environmental Management Act in a more general and qualitative form. As part of the VIEW Project, a seminar and two workshops were organized, at which applicants, officials from various bodies involved in the licensing process and representatives of environmental groups had an opportunity to exchange experiences and information. The views expressed at these gatherings and other informally conveyed information more or less endorse the impression of the licensing system obtained from the Employers' Unions survey.

The main points to emerge are as follows:

- Applicants are unhappy about the amount of information required from them and would like to see standard forms introduced.
- There is a trend away from detailed, rigid formulation of conditions, towards more open and flexible licensing (although little evidence is yet available, especially regarding the enforcement of more open licences).
- The authorities are still learning to utilize the wider scope provided by the Environmental Management Act and the option of imposing conditions relating to energy, waste and traffic as well as 'traditional' pollutants. The learning process, which is sometimes hampered by lack of information, is still at an early stage and involves a significant element of trial and error.
- The same can be said regarding the application of the Alara principle.
- In practice, it is hard to base emission standards and licence conditions on quality standards. In other words, it is difficult to use existing quality standards or quality targets as assessment criteria, as the Environmental Management Act requires.

- Where larger installations are concerned, the Environmental Management Act has to some extent been successful in promoting the inclusion of targets in licencing conditions, rather than specific measures, but it has not had this effect where smaller installations are concerned.

V Implementation in the Netherlands

In the last part of this Chapter I will try to explore the implications of implementing the IPPC Directive, as most recently drafted, in the Netherlands. At first glance one is inclined to think that the Environmental Management Act's integrated licensing system meets the requirements of the Directive to a very large extent. This system is based on an integrated approach as required in Article 6 of the Directive. One can also say that application of the Alara principle in the Netherlands meets the Directive's requirement regarding the use of the best available techniques, in practice at least. In general I think that we may say that we also meet the Directive's requirements concerning the openness of licensing procedures and access to information. However, closer consideration gives rise to serious doubt on some issues.

First, and fundamentally, it is not clear whether the word 'installation' used in the Directive means quite the same as the Dutch word *inrichting*, which is sometimes translated as 'establishment'. Looking at Annex I, which for instance indicates that mineral oil sites and refineries are installations, my inclination is to say that the Dutch word *inrichting* is synonymous with the word 'installation', as used in the Directive. On the other hand, Article 2.8 states that a permit may cover one or more installations or parts of an installation at the same site, while in the Netherlands a licence covers a single installation or establishment, and any extensions or additions require additional licences. If 'installation' is considered to be synonymous with 'establishment', Article 6 of the Directive presents the Netherlands with a serious problem. It mentions the procedure and conditions for granting a permit, which have to be fully coordinated where more than one competent (licensing) authority is involved. How should one interpret this article? Does it mean that there should only be one permit for each installation? Earlier in the development of the Directive, this was certainly the intention, but is it still? As far as I can see, the Directive does not allow for the possibility of several permits being issued to cover one installation. If this is correct, we will have to adapt the Dutch licensing system, under which discharges into surface waters are not covered by the integrated environmental licence, but by a separate licence issued under the Pollution of Surface Waters Act, as I explained before. This licence is coordinated with, but not incorporated within, the integrated environmental licence.

I believe Jit Peeters mentions in his article that Article 8.6 allows scope for the use of general rules to impose certain requirements regarding certain categories of installation. However, 'certain' is the operative word here; full replacement of a licence by general rules would not be possible for the installations listed in Annex I. All the same, I do not expect that this provision will cause any real trouble in the Netherlands.

The Annex I installations are the bigger ones, for which full replacement of licences by general rules is not under consideration in the Netherlands.

The requirement in Article 7, that permits must include details of the arrangements made for air, water and land protection also begs one or two questions. Not least, what does the requirement mean? Will it restrict movement towards more open licences? Furthermore I am not quite sure about the preference in Article 8 for emission limit values, but where appropriate these values may be replaced by technical measures. On the other hand, superficial reading of Article 8.2a suggests that the use of any technique or specific technology should not be prescribed, as was suggested yesterday. I am inclined to read 'without prescribing' in connection with best available techniques which should remain a general principle and should not be reduced to one particular technique or technology.

I am reasonably sure that the licensing system and practical arrangements in the Netherlands do not fully meet the requirements of Article 2.a under f – measures taken upon definitive cessation of the activity – or of Article 8.4 – measures relating to extraordinary or emergency conditions. I am also sure that Article 9 can only be applied in a restricted number of cases in which the relation between the quality standards and the required additional measures is quite clear.

So a close examination of the Directive's permit system raises a number of questions and certainly suggests that we will have to adapt the relatively new system of integrated environmental licensing introduced by the Environmental Management Act.

Economic Appraisal

Integrated Pollution Prevention and Control: an Economic Appraisal

M.G. Faure and J.G.J. Lefevere[*]

I Introduction[1]

On 14 September 1993 the European Commission formally tabled the proposal for a Council Framework Directive on Integrated Pollution Prevention and Control (IPPC).[2] The concept of Integrated Pollution Control, already announced in the fifth environmental action programme, where it was indicated as one of the priority fields of action,[3] fits perfectly in the 'new' European Community (EC) legislative approach towards integration and consolidation of environmental legislation.[4] The Commission proposal went through various amendments during the negotiation process.[5] On 22 June 1995 the Council unanimously adopted the heavily amended common position.[6]

* M.G. Faure: professor of Comparative and International Environ-mental at Maastricht University.
J.G.J. Lefevere: formerly a lecturer in environmental law at Maastricht University; currently staff lawyer, FIELD, London.

1 A shortened version of this Chapter was published in EELR (1996), pp. 112–122, under the title 'The Draft Directive on Integrated Pollution Prevention and Control: An Economic Perspective'. The authors wish to thank Jan Bongaerts for his useful comments on an earlier version of this paper.
2 COM (93) 423 final, Brussels 14 September 1993, OJ 17 November 1993, C 311.
3 Towards sustainability, A European Community programme of policy and action in relation to the environment and sustainable development, OJ 17 May 1993, C 138, at p. 95.
4 See, *inter alia*, the Council Resolution of 8 June 1993 on the quality of drafting of Community legislation, OJ 17 June 1993, C 166/1.
5 See COM (93) 88 final, Brussels 15 May 1995.
6 OJ 25 March 1996, C 087, p. 8.

Given the EC's history of fragmented environmental legislation[7] the development towards integration is certainly an improvement. The IPPC proposal is a first attempt to integrate some of the sectoral legislation into one decision-making procedure. To this effect the IPPC proposal introduces a new system of environmental standard setting. In this new system the permit plays a central role. Integrated permits are required for polluting activities (if they fall under Annex I of the Directive). These permits have to contain emission standards which have to be set on the basis of Best Available Technology (BAT) with reference to Environmental Quality Standards. In this Chapter we will try to give an economic analysis of the method of integration opted for in the IPPC proposal. We have chosen an economic approach in order to highlight the possible effects of the combination of instruments suggested in the proposal.

We will first give a short description of the contents of the proposal (II). In the economic analysis of law much attention has been paid to the various legal instruments that could be used to reach the economic goal of internalizing the externality caused by the pollution. In section III we will give a brief overview of these approaches. After this, we will indicate some economic benefits of the concept of integration (IV). Then we will address the question whether the subsidiarity principle is fully taken into account in the proposed Directive, given the economic interpretation of this principle (V). The question then arises whether it is justified to attribute such a central role to the permit, as the proposal does (VI) and the economic meaning of the concept of BAT will be analyzed (VII). Obviously there are many more aspects of the proposal that would merit an economic analysis. Within the scope of this Chapter we will, however, limit ourselves to a discussion of the above mentioned topics. Our contribution will finally provide a few concluding remarks (VIII).

II The Proposal for the IPPC Directive

The goal of the proposal for the Integration of Pollution Prevention and Control Directive is to integrate important aspects of the prevention of pollution and the control of pollution. To this end, the proposal contains both procedural and substantive integration requirements.

As regards the substantive integration requirements the proposal prescribes *permitting* as the obligatory instrument to control emissions from industrial installations (Art. 4). The proposal specifies certain conditions that have to be included in the permit, which have to include at least emission limit values for the substances and preparations listed in Annex III of the Directive (Art. 9). These emission limits have to be set on the basis

7 The fragmentation of European Environmental legislation has been widely criticized. See, *inter alia*, L. Krämer, *Focus on European Environmental Law*, Sweet & Maxwell, London (1992), p. 154; J.H. Jans, *Europees Milieurecht in Nederland*, Woltersgroep, Groningen (1995), p. 163; J.H. Jans, 'De kwaliteit van de Europese milieuwetgeving', *SEW* (1990), p. 503; Discussion of the Report of the Koopmans Commission, in 'Kwaliteitseisen voor Europese regelgeving', *NJB* (4 August 1995), p. 1059.

of BAT (Art. 9(4)), although without prescribing a specific technique. Furthermore, the emission limit values contained in the permit have to be set by reference to environmental quality standards (EQS) (Art. 10).

As regards the procedural requirements, the proposal specifies the information that the application for a permit has to contain (Art. 6). Regarding the decision on the application, Member States are required to coordinate the procedure and the conditions for the grant of the permit (Art. 7). After the granting of the permit, the authority has to ensure a periodical reconsideration of the permit (Art. 13). Member States have to ensure compliance with the permit (Art. 14) and make available to the public the application, the decision of the authority and the permit itself, as well as the results of the monitoring (Art. 15).

In this Chapter we will focus on the regulatory system introduced in the IPPC proposal, more specifically the obligation to establish a *permit system* and the obligation to include in this permit *emission standards*, set on the basis of BAT and with reference to EQS.

III Economic Approach of Pollution Control

In this part of the Chapter we shall demonstrate how economic methodology can be used to provide a critical analysis of the proposed IPPC proposal. We will therefore first explain the difference between public interest and private interest analysis, which is highly relevant when discussing the choice of legal instruments to be used for pollution control (III.1). We will then discuss how an internalization of externalities through efficient standard setting can be reached, looking both at public interest and private interest aspects (III.2).

The reader should keep in mind that we do not claim that economics can provide a final answer to all questions relating to the use of the law to control the pollution problem. The criterion by which the economic analysis judges certain legal rules is obviously an economic criterion such as efficiency or wealth maximization. On the normative level, the policy maker might, however, wish to take into account other values as well when finally deciding upon the level of government action and the policy instruments to be used. This will be especially important when discussing the economic analysis of the subsidiarity principle. However, even if one accepts that values other than efficiency will play a role at the policy level, the economic analysis is still highly useful, since it can make the costs and benefits of certain policy choices apparent.

1 *Methodology: Public Versus Private Interest Analysis*

A Controlling the Externality

In the economics of environmental law, environmental pollution problems are generally described as being *externalities*. Firms do not take the environment and environmental damage into account in the economic decision-making process. Costs of environmental pollution do therefore not play a role. In order to integrate environmental concerns in the decision-making, these concerns have to be *internalized*.[8] Instruments have to be found which require the integration of environmental considerations and costs in the decision-making.

Coase taught that in a primitive world, without transaction costs the internalization would take place even without the force of law, simply because potential victims and the potential polluter would come to the efficient solution through negotiations.[9] However, in most large-scale pollution cases the number of victims will be too large and transaction costs will be prohibitive. Hence, some kind of intervention on the part of the legal system is needed to effectuate the internalization. The public interest analysis of legal instruments then provides an overview of the various types of legal instruments that can be used to reach this internalization.

Another question addressed in this literature is what is the practical meaning of internalization? To what level should the potentially polluting company take the pollution into account? The answer is that a marginal cost/marginal benefit (mc=mb) test should be applied, whereby the marginal costs of further pollution abatement are compared with a further reduction in environmental damage. This weighing of marginal costs and marginal benefits can be translated into an environmental standard, which usually determines the quality and quantity of the emissions that are allowed into the environment: the so-called emission limit values of the IPPC proposal. Public finance theorists and cost benefit analysts have for a long time advocated public policies to control externalities on a marginal cost equalling marginal benefit basis. This literature on policy analysis[10] builds on the earlier work of Pigou.[11] When determining the efficient standard this public interest literature can provide an insight into the costs and benefits of various legal instruments that could be used to give the potential polluter an incentive to follow the efficient standard.

B Interest Group Aspects

Before going into more detail as regards this literature it is worth mentioning that an alternative approach to the standard-setting problem exists, where attention is paid to the influence of private interest groups. Indeed, the 'public interest' literature would examine the choice between, e.g. liability rules and government regulation, by focusing on, e.g. the question whether administrative authorities have better information than

---

8 For an overview of the economics of environmental policy see e.g. W. Baumol and W. Oates, *The Theory of Environmental Policy*, Prentice Hall New Yersey, p. 133–144.

9 R.H. Coase, 'The Problem of Social Cost', *Journal of Law and Economics* (1960), pp. 1–44.

10 L. Friedman, *Microeconomic Policy Analysis*, McGraw-Hill, New York (1984); E. Stokey and R. Zeckhauser, *A Primer for Policy Analysis*, Norton, New York (1978) and for an overview S. Rose-Ackerman, *Rethinking the Progressive Agenda, the Reform of the American Regulatory State*, The Free Press, New York (1992), pp. 14–19.

11 A.C. Pigou, *A Study in Public Finance*, MacMillan, London (1951).

the parties in the accident setting.[12] In that case it is assumed, however, that the decision to introduce regulation will be made with the view of achieving allocative efficiency and that standards are also set taking into account the marginal cost/marginal benefit test. However, reality is often very much different. Government does not only pass a regulation if it is considered to be efficient according to the economic criteria for regulation. In many cases the contents of the regulation can be said to be inefficient. Economists of the 'public choice' school give reasons for these inefficiencies. In their view the politician himself is also a *homo economicus*, who will have a preference for the introduction of regulation which benefits private interest groups that support him. Regulation is in that view not considered as always being in the public interest, but as a product of supply and demand on a political market. On the demand side various interest groups can be discovered who wish favourable legislation, e.g. limiting market entry for competitors. On the supply side are the politicians who wish to maximize their own utility by favouring interest groups in exchange for political support, e.g. during the election campaign.[13]

The reason why a wealth transfer to certain interest groups can take place (the so-called rent seeking) can be found in two well-known economic phenomena: information and transaction costs. The information costs are incurred by the public at large in discovering the effect of the regulation. Because the wealth transferred to the interest group is derived from the public at large, one individual will usually not realize that rent-seeking by an interest group has taken place. The transaction costs are those incurred by the group that wishes to defend a common interest. Only when a group is relatively small, well-organized and single-issue oriented, will it have low transaction costs and be able to successfully lobby the legislator.[14]

For a long time it was thought that safety regulation, such as environmental legislation, should not be considered as a result of lobbying by interest groups, since this type of regulation is always issued in the general interest. However, Maloney and McCormick have shown that regulation is sometimes used as an instrument of rent seeking when it realizes a wealth transfer to certain interest groups. They argue that if industry realizes that regulation is unavoidable, they will cooperate with the politicians in the development of the regulation and try to change the contents to its advantage.[15] A classic example is the introduction of a so-called 'grandfather-clause', which stipulates that new regulation will not be applicable to firms which are already

12 See e.g. S. Shavell, 'Liability for Harm Versus Regulation of Safety', *Journal of Legal Studies* (1984), pp. 357–374.

13 See J. Buchanan, 'Rent Seeking and Profit Seeking', in J. Buchanan, R. Tollison and G. Tullock, eds., *Toward a Theory of the Rent Seeking Society*, A & M University Press, Texas (1980) and J. Buchanan and G. Tullock, *The Calculus of Consent*, University of Michigan Press (1962).

14 See for the conditions of successful lobbying generally M. Olson, *The Logic of Collective Actions*, Harvard University Press, Cambridge (1971).

15 See M. Maloney and R. McCormick, 'A Positive Theory of Environmental Quality Regulation', *Journal of Law and Economics* (1982), pp. 99–123.

on the market. Hence, the regulation creates a barrier to market entry for newcomers and protects the industry already on the market.

The influence of private interest groups cannot only be felt on the level of the legislator. Political scientists have also developed the capture theory, arguing that inevitably administrative agencies regulating the industry will be 'captured' by the industry they are supposed to regulate, *inter alia* because the agency is often dependent on the information provided by the industry.[16] This capture theory is obviously of importance since it can explain why under certain circumstances there is the danger that an administrative agency will not set standards in the public interest, but to protect the industry.

We believe that a combination of this public interest approach and the private interest group theory can be used not only to understand the existence of the current environmental regulation, but also to develop a general framework of environmental policy instruments. If one is aware of the conditions under which an interest group will be successful, as *inter alia* set out by Olson, one can also think of policy devices to cope with the possible influence of interest groups in order to align public policy more with the efficient solution.

2 *Internalizing Externalities through Efficient Standards*

As we have explained under III.1, in principle environmental standards should be set, based on a weighing of marginal costs and marginal benefits. The advantage of this approach is that it can provide an answer to the question at what level standards should be set. Society does not wish to prevent environmental pollution at all costs. Investing in highly expensive cleaning technology that would not additionally reduce environmental damage would mean a waste of scarce resources. Therefore, in answering the question what kind of abatement measures can be required from industry, environmental economics argues that the costs of these measures should be weighed against their potential benefit in the reduction of environmental damage.[17] In practice often various environmental techniques are available at different prices and all lead to different levels of accident reduction. In the ideal world one would choose the efficient environmental technique, being the one where the marginal costs equal the marginal benefits in the reduction of environmental damage. Usually this balancing process corresponds with a certain emission limit value, which is translated into a legal standard. It is important to stress that from an economic point of view legal instruments will have to be used to prevent the environmental damage in the first place. Preventing here means using abatement techniques to efficiently reduce the environmental damage up to the point where mc=mb.

Various legal instruments can be used to reach this goal. One possibility is to use

16 See on the capture theory A.F. Bentley, *The Process of Government*, Chicago (1908) and M. Utton, *The Economics of Regulating Industry*, Basil Blackwell, London (1986), pp. 26–27.

17 See W. Baumol and W. Oates, *The Theory of Environmental Policy*, Prentice Hall, New Jersey, pp. 133–144.

the liability system. By using liability law a potential polluter could in principle be given an incentive not to pollute or to invest in cleaning equipment of which the marginal costs are equal to the marginal benefits in the reduction of environmental damage. An efficiently applied liability law could give incentives to potential polluters to follow the efficiency standard.[18]

A second possibility, much advocated by economists, is to use financial means to force the polluter to follow the optimal standard. This could take the form of a so-called pollution rights model, with which there have been experiments in the United States. Another, more classic financial instrument could be the Pigovian tax. The idea is to set the tax equal to the amount of pollution caused, in order to provide an incentive for avoiding the pollution and thus to have to pay the tax. These kinds of regulatory taxes or levies can now be found in many areas of environmental law in various European States.

A third possibility is to fix the efficient environmental standard *ex ante* through regulation. In an optimal world, where the regulator sets environmental standards in the public interest, the regulator will take into account the marginal costs of more stringent environmental standards and balance these against the marginal benefits in additional accident reduction. One advantage of the regulatory solution is that the administrative authorities might possess better information on the optimal environmental standard than the parties in a private accident setting. However, with standard setting through regulation there is always the risk of capturing through interest groups, a situation which does not exist with liability rules.

Also within the administrative solution, various options are open. The legislator could fix the standards *ex ante* in binding general rules. Another extreme would be that standards are set at the local level in individual permits to be delivered by administrative agencies. Some mix between the two is possible as well. Finally, there is the choice between emission limit values and quality standards.

Obviously, the legal system will not blindly choose for one of the many just mentioned legal instruments. In most environmental law systems one will find a combination of direct regulation, environmental liability and financial incentives via a taxing system. Indeed, in most cases such an optimal mix of legal instruments is best suited to counter the pollution problem, given the inherent flaws that each of these legal instruments has. To some extent this can be compensated by a combination of the various legal instruments available.

Without going into detail concerning this choice of instruments problem within the scope of this Chapter, let us briefly address some issues that a regulator, wishing to use legal instruments in an efficient way, might want to consider. A first point to remember is that the environmental standard required in the legal system (either as a due care level in tort law or as an emission limit value in a permit) will be influenced by the developments in technology. Since technology can change the efficient standard will

---

18 This has been shown in the abundant literature on the economics of accident law. See e.g. S. Shavell, 'Strict Liability Versus Negligence', *Journal of Legal Studies* (1980), pp. 1–25 and S. Shavell, *Economic Analysis of Accident Law*, Harvard University Press, Cambridge (1987).

change over time as well. There is clearly a disadvantage in merely using permits, since permit conditions cannot rapidly be adapted to the technological evolution. This might hence be an argument for using the tort system in addition to a permit system. Second, when standards are set via regulation, devices have to be developed to counter the possibilities of lobbying. This once more is an argument for combining tort law with a regulatory solution. The risks of lobbying also have to be taken into account when the level of government that should set the standards is chosen. In principle a local agency might have the best information on the local environmental conditions that have to be taken into account when standards are set. However, local authorities might be more sensitive to political influence by the industrial community in which they operate than a national regulator, who is further away from social-economic influences. However, it might be possible to enjoy the benefit of the better information of the local agency if capturing is discouraged through a highly transparent licensing process (e.g. by granting rights to citizens and environmental groups as well) and if the decision to grant a permit were given to a non-political institution, preferably to a technical committee of experts.[19]

If, however, emission limit values are in principle set at the national level one should also be cautious with decrees setting emission limit values for an entire industry. A Belgian example illustrates how things can go wrong. The Belgian Surface Water Protection Act had a system of emission limit values that could be fixed for every industry sector separately. This inevitably opens the possibilities for lobbying, since every industrial group will try to achieve favourable emission limit values for its members. At one point there were more than 40 Royal Decrees with emission limit values for certain industry branches that were far milder than the general emission standards. Even for the producers of titanium dioxide a Royal Decree of 4 August 1986 came into being, provided that the pH could not be lower than 1.5, whereas the general rule was that pH had to vary between 6.5 and 9. Apparently the industry lobbied successfully for separate emission limit values. Transaction costs were apparently quite low, since at the time of that Royal Decree there were only two producers of titanium dioxide in Belgium.[20] The conditions for successful lobbying were therefore fulfilled. The Royal Decree, moreover, also contained a grandfather-clause, containing more stringent conditions for newcomers on the market.[21] Grandfather-clauses can, by the way, also be found in many European environmental directives. This is the case, for example, as regards Articles 4 and 6 of the Titanium

19 In Flanders environmental permits are still granted by political institutions at local or provincial level. It is argued that this has the advantage that they can rightly balance the socio-economic interests at stake; the obvious disadvantage is, however, that they are very much subjected to lobbying forces which does not guarantee an efficient outcome.

20 So L. Lavrysen, 'Enkele recente ontwikkelingen van het milieurecht in het geregionaliseerde België', *Milieu en Recht* (1986), p. 300.

21 For further details see M. Faure and R. van den Bergh, *Objectieve aansprakelijkheid, verplichte verzekering en veiligheidsregulering*, Maklu Uitgevers, Antwerp (1989), pp. 163–164.

Dioxide Directive[22] which shows the success of that industry branch as a rent seeker. In sum, if one wishes to set environmental standards in the public interest, one should be very cautious with grandfather-clauses. A transition period may be acceptable under certain circumstances, provided that it is very short and eventually forces the existing firms to comply with the same conditions as newcomers on the market.

Finally, the best way to counter the risk of lobbying is to decrease the information costs to the public at large and to dismantle the rent-seeking efforts. Therefore the licensing process should be as transparent as possible. The same should be true, however, for every procedure in which emission limit values are set, both on the national and on the European level. If a regulator is forced to show that a certain chosen standard corresponds with the marginal cost/marginal benefit test there is the greatest likelihood that standards might eventually align with the efficient solution.

IV Economic Benefits of Integration

The concept of integrated pollution control itself fits in with the economic idea that the legal instruments should be shaped in such a way that the externalities caused by the environmental pollution are internalized as much as possible. A classic problem with a non-integrated pollution control system is that it does not take into account the possible interchangeability of emissions. Because of the in practice very fragmented and sectoral pollution control, once environmental regulation is established in one sector, one can often see a shifting of the pollution to other, less controlled or regulated sectors. Strict controls on emissions into water, for example, have frequently led to installing water purification equipment which in its turn produced sewage sludge which was again dumped on land. Instead of internalizing the costs of environmental pollution these costs only became an externality in another sector. Internalizing the externalities, however, refers to all possible forms of emissions from a certain entity, therefore integrating these different concerns in the decision-making process.

Preventing the shifting of emissions has indeed been advanced as one of the central goals of the proposed IPPC proposal. The original proposal explicitly mentioned under the specific operation objectives that the Directive aims to prevent pollution emissions from larger installations 'so that pollution problems are solved rather than transferred from one part of the environment to another'.[23] The considerations preceding the proposed Directive mention that 'separate approaches to controlling emissions into the air, the water or the land alone may encourage the shifting of pollution across environmental media rather than protecting the environment as a whole'.[24]

The advantage of an integrated approach is therefore that all possible emissions, whether in the air, liquid or solid, can be integrated into one decision-making

22 Directive 89/428 of 21 June 1989, OJ 1989 L 201/56.

23 COM (93) 423 final, p. 25.

24 COM (93) 423 final, p. 30.

procedure as a result of which emission limit values are set in such a way that the industry has no incentives to shift its emissions from one compartment of the environment to the other. Whether the end-result of this procedure will also be efficient, will obviously largely depend upon the procedure and the authorities involved and their ability and willingness to take into account the various possible emissions and set standards in such a way that a shifting is indeed prevented. A second condition is that not only limit values for the various types of emissions are set in a balanced manner, but that also enforceability of standards can be guaranteed. Article 13 of the proposed Directive stipulates that the Member States shall take the necessary measures to ensure that the conditions of the permit are complied with by the operator. An integrated pollution control system will obviously only be successful if not only the standard setting process is integrated, but also the enforcement system. Since the enforcement and implementation of the proposal is left to the Member States, the true value of the integration will depend on their approach.

An integrated approach towards the pollution problem is not only desirable because of the risk of transfer of emissions, but also to enable an efficient calculation of the costs and benefits. Indeed, if one agrees that the mc=mb test should play a role when emission limit values are set, it is desirable that a standard setting authority takes simultaneously into account all emissions in order to be able to decide what the optimal standard to reach the target quality should be.

The concept of an integration of pollution control systems can also be defended by referring to the reduction of administrative costs this will probably bring. A separate licensing procedure for every possible emission from a large plant obviously leads to high costs, both for the industry and for the bureaucracies involved. One can therefore expect that industry would welcome an integrated approach. This has the advantage that industry only has to address itself to one counter instead of having to apply for three or more licences before it can start up its activities. The original proposal of the Directive also aimed at a streamlined decision-making process: in principle decisions with respect to a permit had be taken within six months.[25] This deadline was unfortunately dropped in the common position.

In general there can hardly be any criticism of the concept of integrated pollution control itself. Also the notion, which is stressed many times in the explanatory memorandum, that the aim of integrated pollution control is to *prevent* emissions and to minimize emissions to the environment as a whole, fits in with the economic approach.[26] There it has also been stressed that legal instruments should be used to induce a potential polluter to prevent emissions in the first place.

---

25 This was at least the case in Art. 7(2) of the September 1993 version of the proposed Directive, but has unfortunately been deleted in the June 1995 common position.

26 The explanatory memorandum stated with respect to the proposal: 'It is founded upon a preventive approach to pollution control' (COM (93) 423 final, p. 14).

V Economic Analysis of the Subsidiarity Principle

We have seen that from an economic point of view the concept of IPPC is certainly desirable. A second question, however, is whether action to achieve this integration should be taken on Community or national level. In answering this question, one can look at the guiding principle for European Community action, the subsidiarity principle. On the basis of Article 3B(2) of the EC Treaty, the Community shall take action 'only if and in so far as the objectives of the proposed action cannot be sufficiently achieved by the Member States and can therefore, by reason or the scale of effects of the proposed action, be better achieved by the Community'.

In this section we will discuss how this subsidiarity principle can be interpreted in an economic way (1) and, second, what this economic interpretation of the subsidiarity principle will have as an effect for environmental policy in general (2) and the standard setting question in particular (3). These economic principles will then be applied to the problem of integrated pollution control by asking why we need Community action with respect to IPPC (4) and whether the competences for standard setting as laid down in the proposed Directive correspond with the economic principles (5).

1 *Economic Criteria for Centralization*

With respect to the first question, that is whether it is possible to provide economic criteria for Community action, we should point at the recent inauguration address of the Utrecht law and economics Professor, Roger van den Bergh. Under the provocative title 'Adieu Bruxelles?', he shed light on the economic analysis of the subsidiarity principle.[27] Van den Bergh argues that one should start with the theory of Tiebout on the optimal provision of local public goods.[28] Tiebout argues that when people with the same preferences cluster together in communities, competition between local authorities will, under certain restrictive conditions, lead to allocative efficiency. Well-informed citizens will move to the community that provides the services that are best adapted to their personal preferences. Building upon this model, Van den Bergh argues that a similar point could be made in favour of the competition between legislators. Citizens would move (so-called voting with the feet) to the community that provides legislation that best corresponds with their preferences. Van den Bergh therefore argues that, from an economic point of view, decentralization should be the starting point, since competition between legislators will lead to allocative efficiency.

However, there are certain conditions under which Tiebout competition will not work and which can therefore constitute arguments in favour of centralization. The most important argument in favour of centralization is the existence of transboundary

27 R. van den Bergh, *Subsidiariteit rechtseconomisch bekeken. Adieu Bruxelles?*, Gouda Quint, Arnhem (1994); an English summary of this address has been published as 'The Subsidiarity Principle in European Community Law: Some Insights from Law and Economics', *Maastricht Journal of European and Comparative Law* (1994), pp. 337–366.

28 C. Tiebout, 'A Pure Theory of Local Expenditures', *Journal of Political Economy* (1956), p. 416.

externalities. If the externalities that have to be regulated cross national borders, there may be an economics of scale argument to shift powers to a higher legal order that has competence to deal with the externality over a larger territory. A second argument in favour of centralization is the so-called prisoners' dilemmas. This is the risk that 'a race for the bottom' would emerge between countries who would wish to attract foreign investment. As a result of these prisoners' dilemmas, countries would fail to enact or enforce efficient legislation, in other words try to externalize environmental costs.

2 *Centralization of Environmental Policy?*

Both arguments may play a role in favour of centralization at the European level if one applies Van den Bergh's criteria to environmental problems.[29] With respect to the environmental problems, it can certainly be argued that they are often transboundary. A great many of the environmental Directives therefore fit in this first criterion for Community action. In this respect one can for instance mention the Directive dealing with the discharge of dangerous substances into the aquatic environment[30] and the Regulation on the transboundary shipment of waste.[31] In both cases there is some risk that the pollution exceeds the national borders of the individual Member State. The prisoners' dilemma argument is often found as well. There it is disguised as the argument that the creation of equal conditions of competition is necessary for the functioning of the common market. However, this argument is repeated constantly in a stereotype way, but its validity can be questioned. The necessity of creating equal market conditions would only be an argument in favour of centralization if it would be clear that environmental costs would otherwise be considerably different and, moreover, that this difference may lead to dislocation of firms to the Member States with the lowest standards (the so-called race for the bottom). It is argued that pollution control costs are only a minor fraction of the total sales of manufacturing industries.[32] Moreover, recent American empirical literature characterizes the effects of environmental regulations as 'either small, statistically insignificant or not robust to tests of model

29 We cannot go into detail as regards this problem, given the scope of this paper. More information is provided in R. van den Bergh, M. Faure and J. Lefevere, 'The Subsidiarity Principle in European Environmental Law: An Economic Analysis', in E. Eide and R. van den Bergh (eds.), *Law and Economics of the Environment*, Juridisk Forlag, Oslo (1995), pp. 121–166.

30 Directive 76/464, discharge of dangerous substances into the aquatic environment, OJ 1976 L 129/23.

31 Regulation 259/93 of 1 February 1993, OJ 1993 L 30.

32 Repetto argues that pollution control costs average only about 1.5 per cent of the value of the total sales of manufacturing industries (R. Repetto, *Trade and Sustainable Development*, UNEP Environment and Trade Series, Vol. 1, pp. 21–22).

specification'.[33] Obviously one could argue that the stringency of environmental regulations might have some effect on new firms in their decision to locate for the first time. Increasing stringency of environmental regulations will, however, not induce existing firms to relocate. It has been argued that it is highly unlikely that a company would sell its plant, acquire a new site, build a new facility, recruit and train new workers simply to save on pollution control costs, totalling less than 2 per cent of sales.[34] Even in the case of new plants, the stringency of environmental regulation typically only has a minor impact on the location decision. Again the already mentioned research by Jaffe, Peterson, Portney and Stavins showed that other criteria such as the tax level, public services and the unionization of the labour force have a much more significant impact on the location decision than environmental regulation.[35]

This evidence therefore substantially weakens the prisoners' dilemma argument for Community action in the field of environmental law. Even if differences between Member States would exist as far as the stringency of environmental law is concerned, this will generally not lead companies to relocate to 'pollution havens'. Nevertheless, one can often note European environmental law dealing with merely local externalities. One can think of the Drinking Water Directive[36] or the Bathing Water Directive.[37] Given the fact that the prisoners' dilemma argument for centralization is rather weak for environmental problems and that the mentioned Directives apparently do not deal with transboundary externalities, they seem hard to fit into the economic framework.

There is, however, probably a third reason for environmental action and the European level, which has to do with guaranteeing all Europeans a similar environmental quality. Sometimes this is referred to as the protection of the 'European environmental and cultural heritage and human health'.[38] Directives that clearly aim at the protection of this 'European heritage' are probably the Habitat Directive[39] and the Wild Birds Directive.[40] If one would only look at Tiebout's theory of competition between legal orders, this 'European heritage' reason for Community action seems rather weak. In a competitive framework one would indeed argue that citizens of a Member State are free to choose the environmental quality that corresponds with their preferences. If the Portuguese would for example decide that in their scale of

33 A.B. Jaffe, S.R. Peterson, P.R. Portney and R.M. Stavins, 'Environmental Regulation and the Competitiveness of US Manufacturing: What Does the Evidence Tell Us?', *Journal of Economic Literature* (1995), pp. 157–158.

34 R. Repetto, op. cit. (note 32), p. 22.

35 A.B. Jaffe, S.R. Peterson, P.R. Portney and R.M. Stavins, loc. cit. (note 33), pp. 148–150.

36 Directive 80/778 on the Quality of Water for Human Consumption, OJ 1980 L 229/11.

37 Directive 76/160 concerning the Quality of Bathing Water, OJ 1976 L 31/1.

38 The Wild Birds Directive states in its considerations that 'whereas conservation is aimed at the long-term protection and management of natural resources as an integral part of the heritage of the peoples of Europe'.

39 Directive 92/43, OJ 1992 L 206/7.

40 Directive 79/409, OJ 1979 L 103/1.

preferences, environmental quality does not range as high at the moment, there would be little reason from an economic point of view to enact Community legislation protecting e.g. the flora and fauna in the Algarve region.[41] One can, however, base European intervention to guarantee a minimum environmental quality to all its citizens as a necessary consequence of the mentioned wish to realize a 'Europe for citizens'. If the goal of the Union is to create an ever closer union among the people of Europe (as is stated in Art. A of the Treaty), Brussels may argue that it wishes to guarantee all citizens a basic environmental quality.

3 Subsidiarity and the Standard Setting Problem

Let us now address the question of how this economic analysis of the subsidiarity principle, related to environmental problems in general, also applies to the standard setting problem. We have previously argued that a possible option to guarantee a basic environmental quality to all of its citizens, irrespective of their individual preferences, is through setting harmonized quality standards at the European level and differentiated emission limit values at the Member State level.[42] The reason for harmonized quality standards is obvious: if one accepts the idea that Europe should guarantee a basic environmental quality for all of its citizens, the legal instrument by which to reach this goal is the setting of quality standards to which the environmental components, no matter where they are within the union, should in principle correspond. Obviously we are referring to quality standards that set high criteria for the quality to be reached and not merely to relatively lenient standards that would function as a 'red flag' as is sometimes the case in practice.

However, the theory of optimal specificity has taught us that the costs of attaining a certain level of environmental protection may well vary with location-specific circumstances.[43] In principle, the differentiation of emission standards should be carried through as long as the advantages of further differentiation outweigh the administrative and information costs that are incurred with a highly detailed standard setting process. The basic idea of this proposal is that the different local circumstances in e.g. Greece, Portugal, Denmark or Belgium have as consequence that emission limit values to reach a similar environmental quality may well vary. However, even if one

---

41 W.P.J. Wils, 'Subsidiarity and EC-environmental policy: Taking Peoples' Concerns Seriously', *Journal of Environmental Law* (1994), p. 90, classifies the protection of endangered species as a psychic spill-over.

42 See M. Faure and J.G.J. Lefevere, 'Some Public Interest and Private Interest Aspects of Environmental Standard Setting in the European Community', in D. Mayes (ed.), *The Evolution of Rules for a Single European Market*, European Commission (1995), pp. 320–351.

43 See with respect to this theory of optimal specificity I. Ehrlich and R. Posner, 'An Economic Analysis of Legal Rule Making', *Journal of Legal Studies* (1974), pp. 257–286 and A.I. Ogus, 'Standard Setting for Environmental Protection', in M. Faure, J. Vervaele and A. Weale (eds.), *Environmental Standards in the European Union in an Interdisciplinary Framework*, Maklu, Antwerp (1994), pp. 25–37.

accepts that, from an economic point of view, differentiated emission standards are needed to reach an optimal level of specificity, one could argue that this does not necessarily mean that these differentiated standards should be set by the Member States. Indeed a prisoners' dilemma could still exist when Member States fix the location-specific emission limit value, but we believe that these can be countered by a European control of the end-product, being the quality to be reached.[44] Second, the information necessary to find the optimal level of specificity will be available within the Member States. Provided that the mentioned prisoners' dilemma can be controlled, the differentiated emission limit value should therefore in principle be set by the Member States. Note, however, that there still might be an argument for Community action with respect to emission standards. Zero-standards could, for example, be set if one agrees that certain (carcinogenic) substances will always have a direct negative impact on the quality to be reached and should therefore generally be banned.

4 *Why Community Action with Respect to IPPC?*

A first question that one could ask, looking at the proposed Directive, is why in the first place Community action with respect to integrated pollution control has to be taken. Although the Commission now usually adds a paragraph on subsidiarity, explaining why Community legislation is necessary in the specific area, the justification given by the Commission is traditionally rather weak. This is also the case if one looks at the explanatory memorandum preceding the IPPC proposal in its September 1993 version. The argument was advanced that integrated pollution control is necessary to set standards in a way that emissions will be least damaging for the environment as a whole. The Directive is intended to provide for an EC framework for pollution control.[45] Although we can of course support this argument for integrated pollution control, in no way does it explain why this justifies Community action. If one agrees that an integrated pollution control system is efficient, why not let the Member States

44 The Commission and the ECJ have in recent cases shown a willingness to focus on practical implementation, whether or not Member States in practice meet the quality standards as set in some of the water Directives, rather than only focusing on the formal implementation of Directives. See *inter alia* the ECJ decision in the *Verviers Drinking Water* case, ECJ 5 July 1990, case C–42/89, *Commission* v. *Belgium*, ECR [1990] I–2821. Also changes in the Treaty (one only has to think of the possibility of a penalty payment) and the recent case law of the European Court of Justice (Francovich liability) have increased the possibilities for an effective enforcement of these quality standards upon the Member States both by citizens and by the Commission, although much improvement is still needed; see J.G.J. Lefevere, 'State Liability for Breaches of Community Law', *EELR* (1996), pp. 237–242; M. Faure and J.G.J. Lefevere, 'Introduction to European Environmental Law', in *Shipping Law Faces Europe: European Policy, Competition and Environment*, Maklu, Antwerp (1995), pp. 93–107; and W. van Gerven, 'Bridging the Gap between Community and National Laws: Towards a Principle of Homogeneity in the Field of Legal Remedies?', *CMLR* (1995), pp. 679–702.

45 See COM (93) 423 final, p. 21.

decide whether or not to install such a regime, given the subsidiarity principle? Several reasons could be advanced why Community action in this respect is indeed necessary, but they cannot be found in the paragraph that is said to be devoted to the subsidiarity principle.

One argument could be taken from Van den Bergh's criteria in favour of centralization. The proposed Directive will only apply to emissions from larger installations. Emissions resulting from these larger installations may be of a transboundary nature. This transboundary character of the externality was an important argument in favour of centralization. However, the pollution itself is not the subject of the IPPC proposal, only the regulatory framework necessary to control the pollution.

Another argument in favour of European intervention could be derived from the 'European Heritage' reason, if quality standards were to be introduced by the IPPC proposal. The IPPC proposal itself does not provide for specific quality standards. It does, however, bring together the decision-making on different sectors of Community environmental legislation, which themselves contain specific emission and quality standards. This brings us to the third – and in this case most valid – argument in favour of Community intervention, which is closely linked to the second argument. Indeed, already a large bulk of Community environmental legislation exists that contains specific standards for specific sectors. The IPPC Directive will now provide for a framework to integrally consider these separate standards. The most valid argument for the Directive is therefore the practical point that it integrates a large body of already existing environmental Community legislation.

5 Competences for Standard Setting

As far as the contents of the IPPC proposal is concerned, it should be mentioned that the Directive itself provides for a division of tasks between the Community and the Member States that seems to fit into the economic framework. As we have mentioned above, three concepts seem to be most relevant in the proposal; Environmental Quality Standards, Best Available Technology (BAT) and Emission Limit Values.

The proposal seems a very important attempt to leave the classic Community approach of EC-wide harmonization of emission limit values. As we have argued above, from an economic point of view it makes more sense to harmonize quality standards and to differentiate the emission limit values that are necessary to reach the specific quality standards. Although not explicitly mentioned, the principle of differentiation of emission limit values seems to have been accepted to a large extent, at least if one reads the explanatory memorandum. When discussing the means for achieving integration, it is *inter alia* mentioned that it will be for each Member State to require emission limit values based on BAT.[46] The point whether or not a uniform regulation is necessary is explicitly discussed in the explanatory memorandum under 2.7. There it is stated:

46 COM (93) 423 final, p. 4.

> The proposal recognizes that the setting of emission limit values can generally best be done at local level, taking into account appropriate environmental conditions. The same standards are not always appropriate at each and every location in the Community.[47]

This statement can of course be supported from an economic point of view. It is furthermore explicitly recognized that standards may differ precisely because areas with a high environmental quality may not need as stringent emission standards. Referring to the emission limit values, the memorandum states that these:

> will be derived from knowledge about the best available techniques to prevent or minimize pollution, although lower standards may be acceptable for some substances, and where the environmental quality of the area is good.[48]

In the explanatory memorandum to the original Commission proposal the clear-cut division in competence regarding the setting of emission standards on national and local level and the setting of quality standards on the European level seems to have been followed. The proposal embraced the idea of harmonized quality standards and differentiated emission limit values.

The text of the original proposal and the later amendments, however, do not seem to follow this same line. Regarding the original proposal one can point at several phrases in the explanatory memorandum, indicating that an exchange of information is necessary in order to assist the Commission in setting future standards at the European level.[49] Apparently the Commission also has a harmonization of emission standards in mind in that respect. Indeed, discussing the necessity of a uniform regulation, the explanatory memorandum to the September 1993 proposal argued that as long as broadly comparable emission limits are set locally across the Community and when the effect on competitiveness is minimal, there may not always be a need for harmonization at Community level. But they continue:

> where, however, standards are very different, so affecting competitiveness, future proposals under the framework of this Directive are much more likely in order to ensure the effective functioning of the internal market.[50]

Here one definitely recognizes the prisoners' dilemmas argument which was, as was mentioned above, rather weakly given the limited influence of environmental measures on the competitiveness of industry. The wish of the Commission to set standards at the European level is also mentioned in the part concerning the follow up and evaluation of the operation. There it is argued that a European standard setting will still be necessary 'where the standards set at Member State level are or become significantly

47 COM (93) 423 final, p. 12.
48 COM (93) 423 final, p. 21.
49 COM (93) 423 final, p. 4.
50 COM (93) 423 final, p. 12.

different'.[51] This is apparently also the reason why Member States should, according to Article 16, send available representative data on the limit values and the best available techniques to the Commission.

The June 1995 version has a new explicit reference to the setting of emission standards at Community level: Article 18. On the basis of this Article the Council shall set emission limit values for the installations in Annex I and the substances in Annex III to the Directive. The new IPPC proposal has therefore lost its framework character by creating a new competence to set emission standards for the substances in its Annex III.

It is noteworthy that with the inclusion of the new Article 18 the proposal seems to respond to the demands of industry to introduce emission standards at Community level, rather than leave the competence to set these at the national or local level. In their reaction to the original proposal industry argued that they would have preferred an immediate agreement on emission limit values at Community level.[52] A strong lobbying effort to this effect was undertaken by *inter alia* German industry. In the summary of the proceedings of the workshop on the IPPC proposal which was held on 5 May 1994 in Bonn, it can be read that the participants at the workshop did not accept the determination of emission limit values at the level of the individual applications, mainly for reasons of legal certainty and the absence of the necessary skills with the local authorities, but also because industry should 'be certain to have to meet the same conditions of competitiveness across Europe'.[53] More particularly Article 9(3) of the original proposal, allowing emission standards lower than BAT, was strongly opposed. This paragraph has indeed been dropped in the most recent June 1995 version.

This industry point of view can be understood from a public choice perspective. Indeed, the industry encounters less influence from environmental interest groups at the European level than at local level. Environmental groups seem to be relatively well organized at the national and regional level where the influence of a certain pollution activity will immediately motivate citizens to associate themselves with environmental groups. Standards at the European level were traditionally proposed by technical committees with very little democratic control. Therefore it seems fair to argue that the influence of industrial lobby groups might be larger at European than at local level. Hence, it should not be surprising that industry preferred an emission standard setting at the central level.[54]

In sum, we believe that the IPPC proposal should explicitly mention that

---

51 COM (930) 423 final, p. 28.

52 COM (93) 423 final, p. 24.

53 Paper printed by the German Bundesumweltministerium: *Umweltpolitik; IPC-Planspiel zur europaeinheitlichen Zulassung von Industrieanlagen*, summary of the Bonn Workshop on 5 May 1994, p. 14.

54 This point is further developed in R. van den Bergh, M. Faure and J.G.J. Lefevere, 'The Subsidiarity Principle in European Environmental Law: An Economic Analysis', loc. cit. (note 29).

environmental quality standards are as far as possible set at Community level. In addition, the Commission should be extremely cautious with further standardization of emission limit values. Even if the standards set at Member State level become 'significantly different', this would not be an argument for centralized European intervention. These 'significant differences' can be explained by location-specific circumstances. From a European point of view, the only question should be whether even diverging emission standards can in the end lead to a common environmental quality within the Union.

VI The Permit as a Central Instrument for Integrated Pollution Control

We shall now focus on the provisions concerning standard setting by means of permits, which is crucial as regards the concept of IPPC as laid down in the proposal. With the entry into force of the IPPC proposal the permit is indeed to become the crucial legal instrument for environmental policy. Article 4 provides that Member States shall take the necessary measures to ensure that no new installation is operated without a permit issued in accordance with this Directive.[55]

The choice of the permit as the most important instrument to be used, is not that self-evident. A whole range of possible legal instruments exist that can control the pollution problem, ranging from economic instruments such as taxes to a variety of legal instruments such as liability rules.[56] Moreover, the public choice theory has taught us that permits might generally have the disadvantage that they can provide possibilities for capture by industry, especially if a large amount of discretion is left to the administrative authority granting the permit. It is indeed questionable why, given the subsidiarity principle, so much weight is put on the permit. In the considerations preceding the Directive, it is merely stated that a system of prior integrated licensing should be established in order to address potential pollution problems before they occur.[57] The question arises, however, whether an integrated approach, which we greatly support, should necessarily be combined with a permit system. Would it not, given subsidiarity, have been preferable to prescribe an integrated approach towards the pollution problem, but then to leave freedom to the Member States to choose the legal instruments they consider most appropriate in order to effectuate this integrated approach. In point 2.7 of the explanatory memorandum, the Commission stated:

55 This is the June 1995 version; in the September 1993 version it is stated that Member States shall take the necessary measures to ensure that 'no installation' shall be operated without a permit; no reference was made that this would only apply to new installations.

56 See for this choice of instruments problem: M.G. Faure and J.G.J. Lefevere, 'Protecting Drinking Water Quality Against Contamination by Pesticides: An Alternative Regulatory Framework', *RECIEL* (1995) Vol. 4:4, pp. 321–326, and S. Shavell, 'Liability for Harm Versus Regulation of Safety', *Journal of Legal Studies* (1984), pp. 357–374.

57 COM (93) 423 final, p. 30.

> The Directive does not attempt to impose one institutional structure for the whole Community – arrangements which are successful in one country, may not be appropriate in another owing to differences, *inter alia*, in national legal and administrative structures.[58]

This being very true, the question arises why the Commission has put the rather limited emphasis on one legal instrument, the permit.

Although the September 1993 version of the Directive seemed to rely exclusively on the permit as a regulatory administrative instrument, the new June 1995 version now pays attention to other obvious administrative instruments as well. In a new Article 3 general principles governing the basic environmental obligations of the operator can be found. These, however, are not obligations that are directly enforceable upon the operator of a certain plant. The new Article 3 provides that it shall be sufficient for the purpose of compliance with this article if Member States ensure that the competent authorities take account of the general principles set out above, when they determine the conditions of the permit. These general principles therefore only guide the standard setting authorities, not the operator of a plant.

An alternative legal instrument often used in national legal systems is setting standards for certain types of activities or industry through generally binding rules. They were absent in the September 1993 version, but the June 1995 version of the IPPC proposal provides that Member States may prescribe certain requirements for certain categories of installations in generally binding rules instead of including them in individual permit conditions, provided that an integrated approach and an equivalent high level of environmental protection as a whole are ensured. As such, generally applicable rules may have the advantage that they can be set by a technical agency or at a more central level, further away from the political and economic influence in the community where the plant is located. However, for these generally binding rules to be efficient, the procedure should be organised in such a way that it is also transparent in order to avoid the risk of lobbying. The question arises, however, whether generally binding rules will play a crucial role for the larger installations to which the IPPC proposal was supposed to apply.

Furthermore, the prevention of pollution at source can be achieved by many legal instruments other than administrative rules like permits. The considerations preceding the Directive seem to suggest that a prior integrated licensing system is necessary to address potential pollution problems before they occur. This goal can, to some extent at least, also be achieved through liability rules,[59] notwithstanding their inherent weaknesses. It would therefore have been interesting if the IPPC proposal would have indicated the relationship with the liability system.

Finally, one should obviously also point to the economic instruments that are often proposed by economists as the most efficient devices to prevent environmental

---

58 COM (93) 423 final, p. 11.

59 This has been shown in the abundant literature on the economics of accident law. See e.g. S. Shavell, 'Strict Liability Versus Negligence', *Journal of Legal Studies* (1980), pp. 1–25 and S. Shavell, *Economic Analysis of Accident Law*, Harvard University Press, Cambridge (1987).

pollution. Although at the national level all kinds of environmental levies and charges are introduced and the feasibility of ecological taxation is examined,[60] the IPPC proposal limits itself to the permit. One can only find reference to economic instruments in one phrase in the considerations preceding the Directive stating that integrated pollution prevention and control 'may be supplemented by the use of economic instruments'.[61] Given the major role that the permit is supposed to play in the pollution control system, it is unclear how the Commission exactly views this supplementary role of economic instruments.

This rather limited view of the Commission is surprising. In the recent European Community programme of policy and action in relation to the environment and sustainable development,[62] entitled *Towards Sustainability*, much attention was given to 'broadening the range of instruments'. A first conclusion of the approach of the IPPC proposal seems to be that instead of broadening the range of instruments, the IPPC proposal has opted for a limitation of the range of instruments. The second conclusion is that there are doubts as to why, in the light of subsidiarity and given the many different approaches in the Member States, the Commission forces all Member States to rely that heavily on the permit system as the main legal instrument to realize environmental policy.

VII Best Available Technology

The concept of Best Available Technology (BAT) is certainly one of the most important aspects of the IPPC proposal. The concept of Best Available Technology Not Entailing Excessive Costs (BATNEEC) has already been in use in several other EC Directives, such as some of the daughter Directives of Directive 76/464 on the discharge of dangerous substances into the aquatic environment[63] and such as the more recent Directives 89/369/EEC and 89/429/EEC on the prevention of pollution from new, and the reduction of air pollution from existing municipal waste incineration plants. The concept of BATNEEC will, however, be removed from these Directives and replaced by the new concept of BAT, as described in the new Article 2(11) of the June 1995 version of the IPPC proposal.

1 BAT as an Efficient Standard

From an economic point of view, environmental standards should in principle be based on a weighing of marginal costs and marginal benefits of pollution and pollution

---

60 A. Paulus, *The Feasibility of Ecological Taxation*, Maklu, Antwerp (1995).

61 COM (93) 423 final, p. 31.

62 See OJ 17 May 1993 C 138/5.

63 *Inter alia*, Directive 83/513/EEC on limit values and quality objectives for cadmium and Directive 84/491/EEC on limit values and quality objectives for discharges of hexachlorocyclohexane.

reduction.[64] In practice this would mean that the technology to be used would be the one where the marginal costs of this technology equal the marginal benefits in reduced environmental damage. If the principle of BAT would allow for such a weighing of costs and benefits it would mean that the economic principles of standard setting would explicitly be recognized in the licensing process.[65]

As we have seen above, in earlier Community legislation reference was made to the notion of BATNEEC. Specifically the 'not entailing excessive costs' seems to refer to the economic marginal cost/marginal benefit test.[66] However, the IPPC proposal does not at first sight seem to refer to NEEC. Although explicit reference to the economic test for standard setting is now missing, one can argue that the marginal cost/marginal benefit test is still at least implicitly present in the current version of the IPPC proposal. The explanatory memorandum to the September 1993 version mentions that although there is no explicit mention of NEEC, 'the definition of BAT in the IPPC Directive takes cost considerations fully into account, for permitting new and existing installations and for upgrading permits. BAT should be considered at the level of the industrial sector and should weigh the environmental benefit of setting standards based on BAT with the cost (or the benefit) to industry of implementing BAT'.[67]

In the September 1993 version a cost/benefit test could indeed be found in Article 2(10) in the definition of the word 'best'. Best means 'most effective in achieving a high level of protection for the environment as a whole, taking into account the potential benefits and costs which may result from action or lack of action'. The explanatory memorandum of September 1993 explicitly stated that 'the definition of "best" reinforces the need to take potential benefits and costs into account in selecting BAT'.[68] The reference to the cost/benefit test in the definition of 'Best' has however been deleted in the June 1995 version. 'Best' now means 'most effective in achieving a high general level of protection of the environment as a whole'.

In Article 2(11) of the new June 1995 IPPC proposal 'best available techniques' signifies:

> the most effective and advanced stage in the development of activities and their methods of operation which indicates the practical suitability of particular techniques for providing in principle the basis for emission limit values designed to prevent and, where it is not practicable, generally to reduce emissions and the impact on the environment as a whole.

---

64 See B. Field, *Environmental Economics, An Introduction*, McGraw-Hill, New York (1994), pp. 45–63 and pp. 133–177.

65 See with respect to these principles also A.I. Ogus, loc. cit. (note 43) and A.I. Ogus, *Regulation, Legal Form and Economic Theory*, Clarendon Press, Oxford (1994), pp. 161–171.

66 For an economic analysis of this BATNEEC notion see M. Faure and M. Ruegg, 'Environmental Standard Setting Through General Principles of Environmental Law', in M. Faure, J. Vervaele and A. Weale (eds.), *Environmental Standards in the European Union in an Interdisciplinary Framework*, Maklu, Antwerp (1994), pp. 43–53.

67 COM (93) 423 final, p. 5.

68 COM (93) 423 final, p. 14.

Reference to some sort of cost/benefit test is now made in the definition of 'available':

> 'available' techniques means those developed on a scale which allows implementation in the relevant industrial sector, under economically and technically viable conditions, taking into consideration the costs and advantages, whether or not the techniques are used or produced inside the Member State in question, as long as they are reasonably accessible to the operator.

Here one can find a clear reference to the fact that costs and benefits should be taken into account when the 'availability' of the technique has to be considered.

In sum, also in the June 1995 version a reference to the cost/benefit test can be found, albeit in the definition of 'available' instead of 'best'. The September 1993 version was preferable in the sense that a comparison of costs and benefits does not determine whether the technique is 'available', but whether this is actually the 'best'. Also the shifting of the cost benefit test from 'best' to 'available' has serious consequences for the level on which the contents of BAT is to be determined, which will be discussed in the next paragraph. Although we have argued that there is certainly a reference to the economic cost/benefit test in the notion of BAT, the precise contents of this concept are still rather vague. Therefore we believe that this notion of BAT, with its reference to the cost/benefit test gives room for an explicit application of the economics of standard setting. The economic literature could therefore be helpful in defining the precise interpretation of BAT.[69]

2 *BAT and Subsidiarity*

A second general point that has to be addressed concerns the relationship between BAT, emission limit values and quality standards in the context of our previous interpretation of subsidiarity and therefore differentiated emission limit values. The IPPC proposal does not seem to be too clear on that point. As we have seen above, the proposal starts from the idea that emission limit values should be set by the Member States.

This corresponds with our economic framework of harmonized quality standards and differentiated emission limit values. Therefore it seems logical that emission limit values should in the first place take into account the environmental quality that has to be reached. Second, if the goal to be reached is clear, being a specific environmental quality (which in our view should be harmonized) BAT can provide information on the costs and benefits of various techniques that meet the targeted quality. However, given different local circumstances the Commission agrees that emission limit values will vary. This, however, is obviously not only the case for the emission limit values, but also for BAT. The marginal cost/marginal benefit comparison will inevitably depend upon local conditions as well. A technology which can be 'best' in Member State A, might not at all be 'best' in Member State B simply because local environmental conditions

69 See in this respect also M. Faure and M. Ruegg, loc. cit. (note 66), p. 52.

are totally different.

In our view, the IPPC proposal is not clear on the exact functioning of BAT in its relation to the emission limit values. On the one hand, it is stated that both BAT and quality standards shall determine emission limit values[70] and that emission limit values in a permit should be based on BAT,[71] but, on the other hand, the proposal seems to accept a differentiation between emission limit values, but not between BAT. This conceptual confusion in the proposal becomes clear if one reads the explanatory memorandum preceding the September 1993 version of the proposal where it discusses the impact of geographical variety.[72] There it is stated that the Directive 'contains a relaxation from the standards of "best available techniques" in certain areas, where the local environmental quality is good and where the emissions would not cause unacceptable environmental effects'. Apparently the proposal contains the notion that BAT are given, regardless of the regional differences. From an economic point of view standard setting would of course take into account the local environmental quality. In that sense adapting an emission limit value to the local environmental quality (in order to reach a similar environmental quality in the whole Union!) is in our view not a deviation from the BAT standard. If BAT really takes into account marginal costs/marginal benefits for that particular region a less costly technology is obviously BAT. Arguing the opposite would mean that the BAT standard should be harmonized which would inevitably lead to inefficiencies precisely because environmental quality and therefore the costs of pollution abatement differ. This confusion also becomes apparent where the Commission argues that the Directive provides for a comprehensive exchange of information and the application of BAT for new installations throughout the Community.[73] Again this apparently starts from the idea that BAT is a universal concept that provides for a similar standard irrespective of local conditions. This also becomes clear if one reads the comment to the notion of 'available'. There it is stated that this refers to techniques which are available anywhere in the world.[74]

If one accepts that emission limit values should be differentiated, in accordance with our interpretation of the subsidiarity principle, the same should of course be true for the standard of BAT which will precisely guide these emission limit values. As we have seen above, the cost/benefit analysis in BAT has been moved from 'Best' in the November 1993 version to 'Available' in the June 1995 version. This move only reinforces the view that the Community is moving towards a European interpretation of BAT, rather than interpreting the principle on a differentiated case by case basis. After all, within a European market the availability of technology hardly differs from region to region. The definition of available even specificly states that availability is independent of 'whether or not the techniques are used or produced inside the

70 COM (93) 423 final, p. 5.
71 Art. 8(2a) of the June 1995 version.
72 COM (93) 423 final, p. 22.
73 COM (93) 423, p. 23.
74 COM (93) 423 final, p. 14.

Member State in question, as long as they are reasonably accessible to the operator'. If one wishes to allow differentiation in the use of BAT, then a local cost/benefit analysis of whether the technology is in a given situation 'Best' should be made.

3 BAT and Technology Requirements

It should also be mentioned that the proposal wisely decided not to prescribe the type of technology the operator has to use. The explanatory memorandum states that 'the operator may use whatever techniques and other technology he likes, as long as they meet the prescribed emission limit values. Under IPPC the role of the competent authority is not to prescribe the technology and other techniques to be used in the operation of an installation, whose choice is best left to the individual operator, thereby keeping operating flexibility with the industrialist'.[75] The current version of Article 9(4) therefore stipulates that the emission limit values shall be based on the best available techniques 'without prescribing the use of any technique or specific technology'.

In the economics of standard setting it has been argued that indeed a disadvantage of so-called specification standards, forcing the operator to employ certain production methods, may have the disadvantage that no incentives are given for innovation; in addition these specification standards may have anti-competitive effects if a rival firm should use alternative processes which attain similar outcomes at no higher costs.[76]

4 The Enforceability of BAT

Finally, the enforceability of BAT as a general principle of environmental policy has to be addressed. As such, the notion that industry should apply best available technology to reduce environmental harm is extremely useful. Certainly if this is interpreted as a mc=mb test it could lead to an efficient outcome. However, the question still arises whether the duty to adopt BAT applies directly to the operator or in principle merely to the permit granting authority. The latter seems to be the case. The way BAT is formulated in the IPPC proposal, it is seen as a duty for the administrative authorities to set emission limit values in the permit conditions based on BAT. BAT does not only play a role when permit conditions are set for new or existing installations, but also when the permit conditions need to be reconsidered or updated. Member States are, according to Article 11, forced to ensure that the competent authority follows or is informed of developments in best available techniques. New techniques might lead to more stringent emission limit values, more particularly if these new techniques can attain a higher reduction of environmental damage at low cost compared to existing technology. Therefore Article 13 stipulates that a reconsideration of permit conditions has to take place in any event where substantial changes in the best available techniques make it possible to reduce emissions

75 COM (93), 423 final, p. 22.
76 A.I. Ogus, loc. cit. (note 43), pp. 28–29.

significantly without imposing excessive costs. But again, this duty to adapt behaviour according to the technological developments merely poses a duty on the administrative authorities to change permit conditions. This is also made clear in the considerations preceding the September 1993 version:[77]

> whereas, because best available techniques will change with time, particularly in the light of technical advances, the competent authority must follow or be informed of such developments, in order to maintain a high level of protection of the environment as a whole.[78]

This approach, formulating BAT merely as a duty for administrative authorities when setting permit conditions, poses serious questions as regards the responsibility of the operator. Why not simply state that if 'substantial changes', (as mentioned in Art. 13(2)) in the best available techniques have occurred that make it possible to significantly reduce emissions without imposing excessive costs, operators are in principle forced to apply this new technology 'within reasonable limits'. The current formulation relies solely on the action of the administrative authorities to adapt permit conditions. It gives the impression that the operator merely has to follow his permit conditions, even though the emission limit values in his permit might have been based on outdated technology. From an economic point of view an operator should indeed adapt its behaviour to changes in technology. Whenever 'substantial changes in the best available techniques make it possible to reduce emissions significantly without imposing excessive costs' it should be the responsibility of the operator to apply these efficient technologies immediately, instead of waiting for a change in its permit conditions. A legal instrument that could attain this goal would be either a general duty for the operator to apply BAT or liability in tort as a 'stop gap' even if the permit conditions were followed.[79] Moreover, a right of standing might be granted to environmental groups to directly enforce this duty of operators to follow BAT. This will also have the advantage that the question of whether or not BAT has been followed by an operator will not merely be addressed by an administrative agency, but also by a judge e.g. in a liability suit. This might pose an important counterweight against lobbying by the industry in setting BAT. One now reads that BAT will be based on whatever 'is affordable for an average operator in the industrial sector concerned'.[80] Industry might have perverse incentives as far as providing information on BAT is concerned. Therefore private action rights that force an operator to follow BAT (in our

---

77 Although Art. 12 has changed considerably in the June 1995 version.

78 COM (93), 423 final, p. 31.

79 See on these complementarities between tort law and regulation: S. Rose-Ackerman, 'Environmental Liability Law', in T.H. Tietenberg (ed.), *Innovation in Environmental Policy, Economic and Legal Aspects of Recent Developments in Environmental Enforcement and Liability*, Elgon, Brookfield (1992), pp. 223–243; S. Rose-Ackerman, *Rethinking the Progressive Agenda. The Reform of the American Regulatory State*, Free Press, New York (1992), pp. 118–131 and S. Rose-Ackerman, 'Regulation and the Law of Torts', *American Economic Review, Papers and Proceedings* (May, 1991), pp. 54–58.

80 COM (93) 423 final, p. 5.

interpretation of being the efficient standard) might give incentives for efficient behaviour even though emission limit values have been set inefficiently as a result of lobbying.[81]

The IPPC proposal also missed the opportunity to formulate BAT as a general principle of environmental policy. The June 1995 version of the IPPC proposal provides for an Article 3 that is misleadingly entitled 'general principles governing the basic obligations of the operator'. One of these principles is that all the appropriate preventive measures should be taken against pollution, in particular through application of the best available techniques. But again this is, contrary to what the title misleadingly suggests, not formulated as an obligation of the operator. Again it is stated that the Member States shall take the necessary measures to provide that the competent authorities ensure that installations are operated in such a way that the principles are respected. Once more it is formulated as a duty for the administrative authorities, not for the operator. Even more tragic is the final sentence of this new Article 3 that shows the real character of this provision:

> It shall be sufficient for the purpose of compliance with this article, if Member States ensure that the competent authorities take account of the general principle set out when they determine the conditions of the permit.

Once more these general principles only bind the competent authorities when setting the permit conditions, but do not provide a direct obligation for the operator of a plant.

VIII Concluding Remarks

From an economic point of view there is little criticism to be formulated against the principle of integrated pollution control in itself. On the contrary, an integrated approach is a better guarantee for the internalization of the externality caused by pollution. In addition, an integrated approach will allow for a more careful marginal cost/marginal benefit analysis when standards are set. Also the reference to costs and benefits in the definition of BAT allows for an economic appraisal of standards. Using economic methodology to approach the IPPC Directive also had the advantage that we could demonstrate some of the benefits of this multidisciplinary approach towards the pollution problem.

The concept of integration is itself extremely useful, but the way in which it has been elaborated in the proposal leaves too much room for discussion. First, the proposal should make the relationship between BAT, emission limit values and quality standards clearer. More specifically, it should clearly state at what level decisions concerning these standards are taken. We would argue that quality standards that will basically influence the other two, will first have to be set at the European level.

81 This has also been defended by M. Faure and M. Ruegg, loc. cit. (note 66), p. 56–57.

Although the preferences of citizens might differ, which would be an economic argument in favour of a differentiation in quality standards, one could argue that Europe should not only be a European *Economic* Community, but also a Europe for citizens. Therefore a similar environmental quality within the Union should be achieved. But given different environmental conditions, emission standards should in principle be differentiated and no further attempt to harmonize these should be made. References to a further harmonization of emission standards in the future should therefore be deleted from this IPPC proposal. Given differences in local environmental quality, the Best Available Technology by which to realize a certain quality will inevitably also differ. This should be accepted by the Commission and therefore no attempt to harmonize BAT should be made either. In this respect the Commission should really respect the subsidiarity principle and abstain from a harmonization of BAT and emission limit values.

Second, the IPPC proposal should not rely too heavily on permits as the sole regulatory instrument for environmental policy. An integrated approach can be achieved through other administrative rules as well but also via economic instruments, which seems to have been totally neglected. Instead of limiting the range of instruments, the Commission should indeed strive for a broadening of the range of instruments. Again, one can question whether the Commission should force Member States to use a permit system, given subsidiarity.

Third, the notion of BAT should be taken seriously. Hence, it should be formulated as an obligation for the operator and not merely as a duty of the permit granting authority to take BAT into account when standards are set. When BAT is formulated as a general duty for operators it can also play a role outside of the administrative permit system.

Council Directive 96/61/EC of 24 September 1996 Concerning Integrated Pollution Prevention and Control

Official Journal 10 October 1996, L 257/26

THE COUNCIL OF THE EUROPEAN UNION,

Having regard to the Treaty establishing the European Community, and in particular Article 130s(1) thereof,

Having regard to the proposal from the Commission,[1]

Having regard to the opinion of the Economic and Social Committee,[2]

Acting in accordance with the procedure laid down in Article 189c of the Treaty,[3]

1. Whereas the objectives and principles of the Community's environment policy, as set out in Article 130r of the Treaty, consist in particular of preventing, reducing and as far as possible eliminating pollution by giving priority to intervention at source and ensuring prudent management of natural resources, in compliance with the 'polluter pays' principle and the principle of pollution prevention;

2. Whereas the Fifth Environmental Action Programme, the broad outline of which was approved by the Council and the Representatives of the Governments of the Member States, meeting within the Council, in the resolution of 1 February 1993

---

1 OJ C 311, 17 November 1993, p. 6 and OJ C 165, 1 July 1995, p. 9.
2 OJ C 195, 18 July 1995, p. 54.
3 Opinion of the European Parliament of 14 December 1994 (OJ C 18, 23 January 1995, p. 96), Council common position of 27 November 1995 (OJ C 87, 25 March 1996, p. 8) and Decision of the European Parliament of 22 May 1996 (OJ C 166, 10 June 1996).

on a Community programme of policy and action in relation to the environment and sustainable development,[4] accords priority to integrated pollution control as an important part of the move towards a more sustainable balance between human activity and socio-economic development, on the one hand, and the resources and regenerative capacity of nature, on the other;

3. Whereas the implementation of an integrated approach to reduce pollution requires action at Community level in order to modify and supplement existing Community legislation concerning the prevention and control of pollution from industrial plants;

4. Whereas Council Directive 84/360/EEC of 28 June 1984 on the combating of air pollution from industrial plants[5] introduced a general framework requiring authorization prior to any operation or substantial modification of industrial installations which may cause air pollution;

5. Whereas Council Directive 76/464/EEC of 4 May 1976 on pollution caused by certain dangerous substances discharged into the aquatic environment of the Community[6] introduced an authorization requirement for the discharge of those substances;

6. Whereas, although Community legislation exists on the combating of air pollution and the prevention or minimization of the discharge of dangerous substances into water, there is no comparable Community legislation aimed at preventing or minimizing emissions into soil;

7. Whereas different approaches to controlling emissions into the air, water or soil separately may encourage the shifting of pollution between the various environmental media rather than protecting the environment as a whole;

8. Whereas the objective of an integrated approach to pollution control is to prevent emissions into air, water or soil wherever this is practicable, taking into account waste management, and, where it is not, to minimize them in order to achieve a high level of protection for the environment as a whole;

9. Whereas this Directive establishes a general framework for integrated pollution prevention and control; whereas it lays down the measures necessary to implement integrated pollution prevention and control in order to achieve a high level of protection for the environment as a whole; whereas application of the principle

---

4 OJ C 138, 17 May 1993, p. 1.

5 OJ L 188, 16 July 1984, p. 20. Directive as last amended by Directive 91/692/EEC (OJ L 377, 31 December 1991, p. 48).

6 OJ L 129, 18 May 1976, p. 23. Directive as last amended by Directive 91/692/EEC.

of sustainable development will be promoted by an integrated approach to pollution control;

10. Whereas the provisions of this Directive apply without prejudice to the provisions of Council Directive 85/337/EEC of 27 June 1985 on the assessment of the effects of public and private projects on the environment;[7] whereas, when information or conclusions obtained further to the application of that Directive have to be taken into consideration for the granting of authorization, this Directive does not affect the implementation of Directive 85/337/EEC;

11. Whereas the necessary steps must be taken by the Member States in order to ensure that the operator of the industrial activities referred to in Annex I is complying with the general principles of certain basic obligations; whereas for that purpose it would suffice for the competent authorities to take those general principles into account when laying down the authorization conditions;

12. Whereas some of the provisions adopted pursuant to this Directive must be applied to existing installations after a fixed period and others as from the date of implementation of this Directive;

13. Whereas, in order to tackle pollution problems more effectively and efficiently, environmental aspects should be taken into consideration by the operator; whereas those aspects should be communicated to the competent authority or authorities so that they can satisfy themselves, before granting a permit, that all appropriate preventive or pollution-control measures have been laid down; whereas very different application procedures may give rise to different levels of environmental protection and public awareness; whereas, therefore, applications for permits under this Directive should include minimum data;

14. Whereas full coordination of the authorization procedure and conditions between competent authorities will make it possible to achieve the highest practicable level of protection for the environment as a whole;

15. Whereas the competent authority or authorities will grant or amend a permit only when integrated environmental protection measures for air, water and land have been laid down;

16. Whereas the permit is to include all necessary measures to fulfil the authorization conditions in order thus to achieve a high level of protection for the environment as a whole; whereas, without prejudice to the authorization procedure, those measures may also be the subject of general binding requirements;

---

7 OJ L 175, 5 July 1985, p. 40.

17. Whereas emission limit values, parameters or equivalent technical measures should be based on the best available techniques, without prescribing the use of one specific technique or technology and taking into consideration the technical characteristics of the installation concerned, its geographical location and local environmental conditions; whereas in all cases the authorization conditions will lay down provisions on minimizing long-distance or transfrontier pollution and ensure a high level of protection for the environment as a whole;

18. Whereas it is for the Member States to determine how the technical characteristics of the installation concerned, its geographical location and local environmental conditions can, where appropriate, be taken into consideration;

19. Whereas, when an environmental quality standard requires more stringent conditions than those that can be achieved by using the best available techniques, supplementary conditions will in particular be required by the permit, without prejudice to other measures that may be taken to comply with the environmental quality standards;

20. Whereas, because best available techniques will change with time, particularly in the light of technical advances, the competent authorities must monitor or be informed of such progress;

21. Whereas, changes to an installation may give rise to pollution; whereas the competent authority or authorities must therefore be notified of any change which might affect the environment; whereas substantial changes to plant must be subject to the granting of prior authorization in accordance with this Directive;

22. Whereas the authorization conditions must be periodically reviewed and if necessary updated; whereas, under certain conditions, they will in any event be re-examined;

23. Whereas, in order to inform the public of the operation of installations and their potential effect on the environment, and in order to ensure the transparency of the licensing process throughout the Community, the public must have access, before any decision is taken, to information relating to applications for permits for new installations or substantial changes and to the permits themselves, their updating and the relevant monitoring data;

24. Whereas the establishment of an inventory of principal emissions and sources responsible may be regarded as an important instrument making it possible in particular to compare pollution activities in the Community; whereas such an inventory will be prepared by the Commission, assisted by a regulatory committee;

25. Whereas the development and exchange of information at Community level about best available techniques will help to redress the technological imbalances in the Community, will promote the worldwide dissemination of limit values and techniques used in the Community and will help the Member States in the efficient implementation of this Directive;

26. Whereas reports on the implementation and effectiveness of this Directive will have to be drawn up regularly;

27. Whereas this Directive is concerned with installations whose potential for pollution, and therefore transfrontier pollution, is significant; whereas transboundary consultation is to be organized where applications relate to the licensing of new installations or substantial changes to installations which are likely to have significant negative environmental effects; whereas the applications relating to such proposals or substantial changes will be available to the public of the Member State likely to be affected;

28. Whereas the need for action may be identified at Community level to lay down emission limit values for certain categories of installation and pollutant covered by this Directive; whereas the Council will set such emission limit values in accordance with the provisions of the Treaty;

29. Whereas the provisions of this Directive apply without prejudice to Community provisions on health and safety at the workplace,

HAS ADOPTED THIS DIRECTIVE:

Article 1

Purpose and scope

The purpose of this Directive is to achieve integrated prevention and control of pollution arising from the activities listed in Annex I. It lays down measures designed to prevent or, where that is not practicable, to reduce emissions in the air, water and land from the abovementioned activities, including measures concerning waste, in order to achieve a high level of protection of the environment taken as a whole, without prejudice to Directive 85/337/EEC and other relevant Community provisions.

Article 2

Definitions

For the purposes of this Directive:

1. 'substance' shall mean any chemical element and its compounds, with the exception of radioactive substances within the meaning of Directive 80/836/Euratom[8] and genetically modified organisms within the meaning of Directive 90/219/EEC[9] and Directive 90/220/EEC;[10]

2. 'pollution' shall mean the direct or indirect introduction as a result of human activity, of substances, vibrations, heat or noise into the air, water or land which may be harmful to human health or the quality of the environment, result in damage to material property, or impair or interfere with amenities and other legitimate uses of the environment;

3. 'installation' shall mean a stationary technical unit where one or more activities listed in Annex I are carried out, and any other directly associated activities which have a technical connection with the activities carried out on that site and which could have an effect on emissions and pollution;

4. 'existing installation' shall mean an installation in operation or, in accordance with legislation existing before the date on which this Directive is brought into effect, an installation authorized or in the view of the competent authority the subject of a full request for authorization, provided that that installation is put into operation no later than one year after the date on which this Directive is brought into effect;

5. 'emission' shall mean the direct or indirect release of substances, vibrations, heat or noise from individual or diffuse sources in the installation into the air, water or land;

---

8 Council Directive 80/836/Euratom of 15 July 1980 amending the Directives laying down the basic safety standards for the health protection of the general public and workers against the dangers of ionizing radiation (OJ L 246, 17 September 1980, p. 1). Directive as amended by Directive 84/467/EEC (OJ L 265, 5 October 1984, p. 4).

9 Council Directive 90/219/EEC of 23 April 1990 on the contained use of genetically modified micro-organisms (OJ L 117, 8 May 1990, p. 1). Directive as amended by Commission Directive 94/51/EC (OJ L 297, 18 November 1994, p. 29).

10 Council Directive 90/220/EEC of 23 April 1990 on the deliberate release into the environment of genetically modified organisms (OJ L 117, 8 May 1990, p. 15). Directive as amended by Commission Directive 94/15/EC (OJ L 103, 22 April 1994, p. 20).

6. 'emission limit values' shall mean the mass, expressed in terms of certain specific parameters, concentration and/or level of an emission, which may not be exceeded during one or more periods of time. Emission limit values may also be laid down for certain groups, families or categories of substances, in particular for those listed in Annex III.

   The emission limit values for substances shall normally apply at the point where the emissions leave the installation, any dilution being disregarded when determining them. With regard to indirect releases into water, the effect of a water treatment plant may be taken into account when determining the emission limit values of the installation involved, provided that an equivalent level is guaranteed for the protection of the environment as a whole and provided this does not lead to higher levels of pollution in the environment, without prejudice to Directive 76/464/EEC or the Directives implementing it;

7. 'environmental quality standard' shall mean the set of requirements which must be fulfilled at a given time by a given environment or particular part thereof, as set out in Community legislation;

8. 'competent authority' shall mean the authority or authorities or bodies responsible under the legal provisions of the Member States for carrying out the obligations arising from this Directive;

9. 'permit' shall mean that part or the whole of a written decision (or several such decisions) granting authorization to operate all or part of an installation, subject to certain conditions which guarantee that the installation complies with the requirements of this Directive. A permit may cover one or more installations or parts of installations on the same site operated by the same operator;

10. (a) 'change in operation' shall mean a change in the nature or functioning, or an extension, of the installation which may have consequences for the environment;

    (b) 'substantial change' shall mean a change in operation which, in the opinion of the competent authority, may have significant negative effects on human beings or the environment;

11. 'best available techniques' shall mean the most effective and advanced stage in the development of activities and their methods of operation which indicate the practical suitability of particular techniques for providing in principle the basis for emission limit values designed to prevent and, where that is not practicable, generally to reduce emissions and the impact on the environment as a whole:

- 'techniques' shall include both the technology used and the way in which the installation is designed, built, maintained, operated and decommissioned,

- 'available' techniques shall mean those developed on a scale which allows implementation in the relevant industrial sector, under economically and technically viable conditions, taking into consideration the costs and advantages, whether or not the techniques are used or produced inside the Member State in question, as long as they are reasonably accessible to the operator,

- 'best' shall mean most effective in achieving a high general level of protection of the environment as a whole.

In determining the best available techniques, special consideration should be given to the items listed in Annex IV;

12. 'operator' shall mean any natural or legal person who operates or controls the installation or, where this is provided for in national legislation, to whom decisive economic power over the technical functioning of the installation has been delegated.

Article 3

General principles governing the basic obligations of the operator

Member States shall take the necessary measures to provide that the competent authorities ensure that installations are operated in such a way that:

(a) all the appropriate preventive measures are taken against pollution, in particular through application of the best available techniques;

(b) no significant pollution is caused;

(c) waste production is avoided in accordance with Council Directive 75/442/EEC of 15 July 1975 on waste;[11] where waste is produced, it is recovered or, where that is technically and economically impossible, it is disposed of while avoiding or reducing any impact on the environment;

(d) energy is used efficiently;

11 OJ L 194, 25 July 1975, p. 39. Directive as last amended by Directive 91/692/EEC (OJ L 377, 31 December 1991, p. 48).

(e) the necessary measures are taken to prevent accidents and limit their consequences;

(f) the necessary measures are taken upon definitive cessation of activities to avoid any pollution risk and return the site of operation to a satisfactory state.

For the purposes of compliance with this Article, it shall be sufficient if Member States ensure that the competent authorities take account of the general principles set out in this Article when they determine the conditions of the permit.

Article 4

Permits for new installations

Member States shall take the necessary measures to ensure that no new installation is operated without a permit issued in accordance with this Directive, without prejudice to the exceptions provided for in Council Directive 88/609/EEC of 24 November 1988 on the limitation of emissions of certain pollutants into the air from large combustion plants.[12]

Article 5

Requirements for the granting of permits for existing installations

1. Member States shall take the necessary measures to ensure that the competent authorities see to it, by means of permits in accordance with Articles 6 and 8 or, as appropriate, by reconsidering and, where necessary, by updating the conditions, that existing installations operate in accordance with the requirements of Articles 3, 7, 9, 10, 13, the first and second indents of 14, and 15(2) not later than eight years after the date on which this Directive is brought into effect, without prejudice to specific Community legislation.

2. Member States shall take the necessary measures to apply the provisions of Articles 1, 2, 11, 12, 14, third indent, 15(1), (3) and (4), 16, 17 and 18(2) to existing installations as from the date on which this Directive is brought into effect.

12 OJ L 336, 7 December 1988, p. 1. Directive as last amended by Directive 90/656/EEC (OJ L 353, 17 December 1990, p. 59).

Article 6

Applications for permits

1. Member States shall take the necessary measures to ensure that an application to the competent authority for a permit includes a description of:

- the installation and its activities,
- the raw and auxiliary materials, other substances and the energy used in or generated by the installation,
- the sources of emissions from the installation,
- the conditions of the site of the installation,
- the nature and quantities of foreseeable emissions from the installation into each medium as well as identification of significant effects of the emissions on the environment,
- the proposed technology and other techniques for preventing or, where this not possible, reducing emissions from the installation,
- where necessary, measures for the prevention and recovery of waste generated by the installation,
- further measures planned to comply with the general principles of the basic obligations of the operator as provided for in Article 3,
- measures planned to monitor emissions into the environment.

An application for a permit shall also include a non-technical summary of the details referred to in the above indents.

2. Where information supplied in accordance with the requirements provided for in Directive 85/337/EEC or a safety report prepared in accordance with Council Directive 82/501/EEC of 24 June 1982 on the major accident hazards of certain industrial activities[13] or other information produced in response to other legislation fulfils any of the requirements of this Article, that information may be included in, or attached to, the application.

---

13 OJ L 230, 5 August 1982, p. 1. Directive as last amended by Directive 91/692/EEC (OJ L 377, 31 December 1991, p. 48).

Article 7

Integrated approach to issuing permits

Member States shall take the measures necessary to ensure that the conditions of, and procedure for the grant of, the permit are fully coordinated where more than one competent authority is involved, in order to guarantee an effective integrated approach by all authorities competent for this procedure.

Article 8

Decisions

Without prejudice to other requirements laid down in national or Community legislation, the competent authority shall grant a permit containing conditions guaranteeing that the installation complies with the requirements of this Directive or, if it does not, shall refuse to grant the permit.

All permits granted and modified permits must include details of the arrangements made for air, water and land protection as referred to in this Directive.

Article 9

Conditions of the permit

1. Member States shall ensure that the permit includes all measures necessary for compliance with the requirements of Articles 3 and 10 for the granting of permits in order to achieve a high level of protection for the environment as a whole by means of protection of the air, water and land.

2. In the case of a new installation or a substantial change where Article 4 of Directive 85/337/EEC applies, any relevant information obtained or conclusion arrived at pursuant to Articles 5, 6 and 7 of that Directive shall be taken into consideration for the purposes of granting the permit.

3. The permit shall include emission limit values for pollutants, in particular, those listed in Annex III, likely to be emitted from the installation concerned in significant quantities, having regard to their nature and their potential to transfer pollution from one medium to another (water, air and land). If necessary, the permit shall include appropriate requirements ensuring protection of the soil and ground water and measures concerning the management of waste generated by the installation. Where appropriate, limit values may be supplemented or replaced by equivalent parameters or technical measures.

For installations under subheading 6.6 in Annex I, emission limit values laid down in accordance with this paragraph shall take into account practical considerations appropriate to these categories of installation.

4. Without prejudice to Article 10, the emission limit values and the equivalent parameters and technical measures referred to in paragraph 3 shall be based on the best available techniques, without prescribing the use of any technique or specific technology, but taking into account the technical characteristics of the installation concerned, its geographical location and the local environmental conditions. In all circumstances, the conditions of the permit shall contain provisions on the minimization of long-distance or transboundary pollution and ensure a high level of protection for the environment as a whole.

5. The permit shall contain suitable release monitoring requirements, specifying measurement methodology and frequency, evaluation procedure and an obligation to supply the competent authority with data required for checking compliance with the permit.

For installations under subheading 6.6 in Annex I, the measures referred to in this paragraph may take account of costs and benefits.

6. The permit shall contain measures relating to conditions other than normal operating conditions. Thus, where there is a risk that the environment may be affected, appropriate provision shall be made for start-up, leaks malfunctions, momentary stoppages and definitive cessation of operations.

The permit may also contain temporary derogations from the requirements of paragraph 4 if a rehabilitation plan approved by the competent authority ensures that these requirements will be met within six months and if the project leads to a reduction of pollution.

7. The permit may contain such other specific conditions for the purposes of this Directive as the Member State or competent authority may think fit.

8. Without prejudice to the obligation to implement a permit procedure pursuant to this Directive, Member States may prescribe certain requirements for certain categories of installations in general binding rules instead of including them in individual permit conditions, provided that an integrated approach and an equivalent high level of environmental protection as a whole are ensured.

Article 10

Best available techniques and environmental quality standards

Where an environmental quality standard requires stricter conditions than those achievable by the use of the best available techniques, additional measures shall in particular be required in the permit, without prejudice to other measures which might be taken to comply with environmental quality standards.

Article 11

Developments in best available techniques

Member States shall ensure that the competent authority follows or is informed of developments in best available techniques.

Article 12

Changes by operators to installations

1. Member States shall take the necessary measures to ensure that the operator informs the competent authorities of any changes planned in the operation of the installation as referred to in Article 2(10)(a). Where appropriate, the competent authorities shall update the permit or the conditions.

2. Member States shall take the necessary measures to ensure that no substantial change in the operation of the installation within the meaning of Article 2(10)(b) planned by the operator is made without a permit issued in accordance with this Directive. The application for a permit and the decision by the competent authority must cover those parts of the installation and those aspects listed in Article 6 that may be affected by the change. The relevant provisions of Articles 3 and 6 to 10 and Article 15(1), (2) and (4) shall apply *mutatis mutandis*.

Article 13

Reconsideration and updating of permit conditions by the competent authority

1. Member States shall take the necessary measures to ensure that competent authorities periodically reconsider and, where necessary, update permit conditions.

2. The reconsideration shall be undertaken in any event where:

- the pollution caused by the installation is of such significance that the existing emission limit values of the permit need to be revised or new such values need to be included in the permit,

- substantial changes in the best available techniques make it possible to reduce emissions significantly without imposing excessive costs,

- the operational safety of the process or activity requires other techniques to be used,

- new provisions of Community or national legislation so dictate.

Article 14

Compliance with permit conditions

Member States shall take the necessary measures to ensure that:

- the conditions of the permit are complied with by the operator when operating the installation,

- the operator regularly informs the competent authority of the results of the monitoring of releases and without delay of any incident or accident significantly affecting the environment,

- operators of installations afford the representatives of the competent authority all necessary assistance to enable them to carry out any inspections within the installation, to take samples and to gather any information necessary for the performance of their duties for the purposes of this Directive.

Article 15

Access to information and public participation in the permit procedure

1. Without prejudice to Council Directive 90/313/EEC of 7 June 1990 on the freedom of access to information on the environment,[14] Member States shall take the necessary measures to ensure that applications for permits for new installations or for substantial changes are made available for an appropriate period of time to the public, to enable it to comment on them before the competent authority reaches its decision.

---

14 OJ L 158, 23 June 1990, p. 56.

That decision, including at least a copy of the permit, and any subsequent updates, must be made available to the public.

2. The results of monitoring of releases as required under the permit conditions referred to in Article 9 and held by the competent authority must be made available to the public.

3. An inventory of the principal emissions and sources responsible shall be published every three years by the Commission on the basis of the data supplied by the Member States. The Commission shall establish the format and particulars needed for the transmission of information in accordance with the procedure laid down in Article 19.

In accordance with the same procedure, the Commission may propose measures to ensure inter-comparability and complementarity between data concerning the inventory of emissions referred to in the first subparagraph and data from other registers and sources of data on emissions.

4. Paragraphs 1, 2 and 3 shall apply subject to the restrictions laid down in Article 3(2) and (3) of Directive 90/313/EEC.

Article 16

Exchange of information

1. With a view to exchanging information, Member States shall take the necessary measures to send the Commission every three years, and for the first time within 18 months of the date on which this Directive is brought into effect, the available representative data on the limit values laid down by specific category of activities in accordance with Annex I and, if appropriate, the best available techniques from which those values are derived in accordance with, in particular, Article 9. On subsequent occasions the data shall be supplemented in accordance with the procedures laid down in paragraph 3 of this Article.

2. The Commission shall organize an exchange of information between Member States and the industries concerned on best available techniques, associated monitoring, and developments in them. Every three years the Commission shall publish the results of the exchanges of information.

3. Reports on the implementation of this Directive and its effectiveness compared with other Community environmental instruments shall be established in accordance with the procedure laid down in Articles 5 and 6 of Directive 91/692/EEC. The first report shall cover the three years following the date on which this present Directive is brought into effect as referred to in Article 21. The Commission shall submit the report to the Council, accompanied by proposals if necessary.

4. Member States shall establish or designate the authority or authorities which are to be responsible for the exchange of information under paragraphs 1, 2 and 3 and shall inform the Commission accordingly.

Article 17

Transboundary effects

1. Where a Member State is aware that the operation of an installation is likely to have significant negative effects on the environment of another Member State, or where a Member State likely to be significantly affected so requests, the Member State in whose territory the application for a permit pursuant to Article 4 or Article 12(2) was submitted shall forward the information provided pursuant to Article 6 to the other Member State at the same time as it makes it available to its own nationals. Such information shall serve as a basis for any consultations necessary in the framework of the bilateral relations between the two Member States on a reciprocal and equivalent basis.

2. Within the framework of their bilateral relations, Member States shall see to it that in the cases referred to in paragraph 1 the applications are also made available for an appropriate period of time to the public of the Member State likely to be affected so that it will have the right to comment on them before the competent authority reaches its decision.

Article 18

Community emission limit values

1. Acting on a proposal from the Commission, the Council will set emission limit values, in accordance with the procedures laid down in the Treaty, for:

- the categories of installations listed in Annex I except for the landfills covered by categories 5.1 and 5.4 of that Annex,
  and
- the polluting substances referred to in Annex III,

for which the need for Community action has been identified, on the basis, in particular, of the exchange of information provided for in Article 16.

2. In the absence of Community emission limit values defined pursuant to this Directive, the relevant emission limit values contained in the Directives referred to in Annex II and in other Community legislation shall be applied as minimum emission limit values pursuant to this Directive for the installations listed in Annex I.

Without prejudice to the requirements of this Directive, the technical requirements applicable for the landfills covered by categories 5.1 and 5.4 of Annex I, shall be fixed by the Council, acting on a proposal by the Commission, in accordance with the procedures laid down in the Treaty.

Article 19

Committee procedure referred to in Article 15(3)

The Commission shall be assisted by a committee composed of the representatives of the Member States and chaired by the representative of the Commission.

The representative of the Commission shall submit to the committee a draft of the measures to be taken. The committee shall deliver its opinion on the draft within a time limit which the chairman may lay down according to the urgency of the matter. The opinion shall be delivered by the majority laid down in Article 148(2) of the Treaty in the case of decisions which the Council is required to adopt on a proposal from the Commission. The votes of the representatives of the Member States within the committee shall be weighted in the manner set out in that Article. The chairman shall not vote.

The Commission shall adopt the measures envisaged if they are in accordance with the opinion of the committee.

If the measures are not in accordance with the opinion of the committee, or if no opinion is delivered, the Commission shall, without delay, submit to the Council a proposal relating to the measures to be taken. The Council shall act by a qualified majority.

If, on the expiry of a period of three months from the date of referral to the Council, the Council has not acted, the proposed measures shall be adopted by the Commission.

Article 20

Transitional provisions

1. The provisions of Directive 84/360/EEC, the provisions of Articles 3, 5, 6(3) and 7(2) of Directive 76/464/EEC and the relevant provisions concerning authorization systems in the Directives listed in Annex II shall apply, without prejudice to the exceptions provided for in Directive 88/609/EEC, to existing installations in respect of activities listed in Annex I until the measures required pursuant to Article 5 of this Directive have been taken by the competent authorities.

2. The relevant provisions concerning authorization systems in the Directives referred to in paragraph 1 shall not apply to installations which are new in respect of the activities listed in Annex I on the date on which this Directive is brought into effect.

3. Directive 84/360/EEC shall be repealed 11 years after the date of entry into force of this Directive.

As soon as the measures provided for in Article 4, 5 or 12 have been taken in respect of an installation, the exception provided for in Article 6(3) of Directive 76/464/EEC shall no longer apply to installations covered by this Directive.

Acting on a proposal from the Commission, the Council shall, where necessary, amend the relevant provisions of the Directives referred to in Annex II in order to adapt them to the requirements of this Directive before the date of repeal of Directive 84/360/EEC, referred to in the first subparagraph.

Article 21

Bringing into effect

1. Member States shall adopt the laws, regulations and administrative provisions necessary to comply with this Directive no later than three years after its entry into force. They shall forthwith inform the Commission thereof.

When Member States adopt these measures, they shall contain a reference to this Directive or shall be accompanied by such reference on the occasion of their official publication. The methods of making such reference shall be laid down by Member States.

2. Member States shall communicate to the Commission the texts of the main provisions of national law which they adopt in the field covered by this Directive.

Article 22

This Directive shall enter into force on the 20th day following its publication.

Article 23

This Directive is addressed to the Member States.
Done at Brussels, 24 September 1996.

For the Council
The President
E. FITZGERALD

ANNEX I

CATEGORIES OF INDUSTRIAL ACTIVITIES REFERRED TO IN ARTICLE 1

1. Installations or parts of installations used for research, development and testing of new products and processes are not covered by this Directive.

2. The threshold values given below generally refer to production capacities or outputs. Where one operator carries out several activities falling under the same subheading in the same installation or on the same site, the capacities of such activities are added together.

1. **Energy industries**

1.1. Combustion installations with a rated thermal input exceeding 50 MW.[15]

1.2. Mineral oil and gas refineries.

1.3. Coke ovens.

1.4. Coal gasification and liquefaction plants.

2. **Production and processing of metals**

2.1. Metal ore (including sulphide ore) roasting or sintering installations.

2.2. Installations for the production of pig iron or steel (primary or secondary fusion) including continuous casting, with a capacity exceeding 2.5 tonnes per hour.

2.3. Installations for the processing of ferrous metals:

(a) hot-rolling mills with a capacity exceeding 20 tonnes of crude steel per hour;

(b) smitheries with hammers the energy of which exceeds 50 kilojoule per hammer, where the calorific power used exceeds 20 MW;

(c) application of protective fused metal coats with an input exceeding 2 tonnes of crude steel per hour.

15 The material requirements of Directive 88/609/EEC for existing installations still apply until 31 December 2003.

2.4. Ferrous metal foundries with a production capacity exceeding 20 tonnes per day.

2.5. Installations:

(a) for the production of non-ferrous crude metals from ore, concentrates or secondary raw materials by metallurgical, chemical or electrolytic processes;

(b) for the smelting, including the alloyage, of non-ferrous metals, including recovered products, (refining, foundry casting, etc.) with a melting capacity exceeding 4 tonnes per day for lead and cadmium or 20 tonnes per day for all other metals.

2.6. Installations for surface treatment of metals and plastic materials using an electrolytic or chemical process where the volume of the treatment vats exceeds 30 m^3.

3. **Mineral industry**

3.1. Installations for the production of cement clinker in rotary kilns with a production capacity exceeding 500 tonnes per day or lime in rotary kilns with a production capacity exceeding 50 tonnes per day or in other furnaces with a production capacity exceeding 50 tonnes per day.

3.2. Installations for the production of asbestos and the manufacture of asbestos-based products.

3.3. Installations for the manufacture of glass including glass fibre with a melting capacity exceeding 20 tonnes per day.

3.4. Installations for melting mineral substances including the production of mineral fibres with a melting capacity exceeding 20 tonnes per day.

3.5. Installations for the manufacture of ceramic products by firing, in particular roofing tiles, bricks, refractory bricks, tiles, stoneware or porcelain, with a production capacity exceeding 75 tonnes per day, and/or with a kiln capacity exceeding 4 m^3 and with a setting density per kiln exceeding 300 kg/m^3.

4. **Chemical industry**

Production within the meaning of the categories of activities contained in this section means the production on an industrial scale by chemical processing of substances or groups of substances listed in Sections 4.1 to 4.6.

4.1. Chemical installations for the production of basic organic chemicals, such as:

(a) simple hydrocarbons (linear or cyclic, saturated or unsaturated, aliphatic or aromatic);

(b) oxygen-containing hydrocarbons such as alcohols, aldehydes, ketones, carboxylic acids, esters, acetates, ethers, peroxides, epoxy resins;

(c) sulphurous hydrocarbons;

(d) nitrogenous hydrocarbons such as amines, amides, nitrous compounds, nitro compounds or nitrate compounds, nitriles, cyanates, isocyanates;

(e) phosphorus-containing hydrocarbons;

(f) halogenic hydrocarbons;

(g) organometallic compounds;

(h) basic plastic materials (polymers synthetic fibres and cellulose-based fibres);

(i) synthetic rubbers;

(j) dyes and pigments;

(k) surface-active agents and surfactants.

4.2. Chemical installations for the production of basic inorganic chemicals, such as:

(a) gases, such as ammonia, chlorine or hydrogen chloride, fluorine or hydrogen fluoride, carbon oxides, sulphur compounds, nitrogen oxides, hydrogen, sulphur dioxide, carbonyl chloride;

(b) acids, such as chromic acid, hydrofluoric acid, phosphoric acid, nitric acid, hydrochloric acid, sulphuric acid, oleum, sulphurous acids;

(c) bases, such as ammonium hydroxide, potassium hydroxide, sodium hydroxide;

(d) salts, such as ammonium chloride, potassium chlorate, potassium carbonate, sodium carbonate, perborate, silver nitrate;

(e) non-metals, metal oxides or other inorganic compounds such as calcium carbide, silicon, silicon carbide;

4.3. Chemical installations for the production of phosphorous-, nitrogen- or potassium-based fertilizers (simple or compound fertilizers).

4.4. Chemical installations for the production of basic plant health products and of biocides.

4.5. Installations using a chemical or biological process for the production of basic pharmaceutial products.

4.6. Chemical installations for the production of explosives.

5. **Waste management**

Without prejudice to Article 11 of Directive 75/442/EEC or Article 3 of Council Directive 91/689/EEC of 12 December 1991 on hazardous waste.[16]

5.1. Installations for the disposal or recovery of hazardous waste as defined in the list referred to in Article 1(4) of Directive 91/689/EEC, as defined in Annexes II A and II B (operations R1, R5, R6, R8 and R9) to Directive 75/442/EEC and in Council Directive 75/439/EEC of 16 June 1975 on the disposal of waste oils,[17] with a capacity exceeding 10 tonnes per day.

5.2. Installations for the incineration of municipal waste as defined in Council Directive 89/369/EEC of 8 June 1989 on the prevention of air pollution from new municipal waste incineration plants[18] and Council Directive 89/429/EEC of 21 June 1989 on the reduction of air pollution from existing municipal waste-incineration plants[19] with a capacity exceeding 3 tonnes per hour.

5.3. Installations for the disposal of non-hazardous waste as defined in Annex II A to Directive 75/442/EEC under headings D8 and D9, with a capacity exceeding 50 tonnes per day.

5.4. Landfills receiving more than 10 tonnes per day or with a total capacity exceeding 25,000 tonnes, excluding landfills of inert waste.

---

16 OJ L 377, 31 December 1991, p. 20. Directive as amended by Directive 94/31/EC (OJ L 168, 2 July 1994, p. 28).

17 OJ L 194, 25 July 1975, p. 23. Directive as last amended by Directive 91/692/EEC (OJ L 377, 31 December 1991, p. 48).

18 OJ L 163, 14 June 1989, p. 32.

19 OJ L 203, 15 June 1989, p. 50.

6. **Other activities**

6.1. Industrial plants for the production of:

(a) pulp from timber or other fibrous materials;

(b) paper and board with a production capacity exceeding 20 tonnes per day.

6.2. Plants for the pre-treatment (operations such as washing, bleaching, mercerization) or dyeing of fibres or textiles where the treatment capacity exceeds 10 tonnes per day.

6.3. Plants for the tanning of hides and skins where the treatment capacity exceeds 12 tonnes of finished products per day.

6.4. (a) Slaughterhouses with a carcase production capacity greater than 50 tonnes per day;

(b) Treatment and processing intended for the production of food products from:

- animal raw materials (other than milk) with a finished product production capacity greater than 75 tonnes per day;
- vegetable raw materials with a finished product production capacity greater than 300 tonnes per day (average value on a quarterly basis);

(c) Treatment and processing of milk, the quantity of milk received being greater than 200 tonnes per day (average value on an annual basis).

6.5. Installations for the disposal or recycling of animal carcases and animal waste with a treatment capacity exceeding 10 tonnes per day.

6.6. Installations for the intensive rearing of poultry or pigs with more than:

(a) 40,000 places for poultry;

(b) 2,000 places for production pigs (over 30 kg); or

(c) 750 places for sows.

6.7. Installations for the surface treatment of substances, objects or products using organic solvents, in particular for dressing, printing, coating, degreasing, waterproofing, sizing, painting, cleaning or impregnating, with a consumption capacity of more than 150 kg per hour or more than 200 tonnes per year.

6.8. Installations for the production of carbon (hard-burnt coal) or electrographite by means of incineration or graphitization.

ANNEX II

LIST OF THE DIRECTIVES REFERRED TO IN ARTICLES 18(2) AND 20

1. Directive 87/217/EEC on the prevention and reduction of environmental pollution by asbestos.

2. Directive 82/176/EEC on limit values and quality objectives for mercury discharges by the chlor-alkali electrolysis industry.

3. Directive 83/513/EEC on limit values and quality objectives for cadmium discharges.

4. Directive 84/156/EEC on limit values and quality objectives for mercury discharges by sectors other than the chlor-alkali electrolysis industry.

5. Directive 84/491/EEC on limit values and quality objectives for discharges of hexachlorocyclohexane.

6. Directive 86/280/EEC on limit values and quality objectives for discharges of certain dangerous substances included in List 1 of the Annex to Directive 76/464/EEC, subsequently amended by Directives 88/347/EEC and 90/415/EEC amending Annex II to Directive 86/280/EEC.

7. Directive 89/369/EEC on the prevention of air pollution from new municipal waste-incineration plants.

8. Directive 89/429/EEC on the reduction of air pollution from existing municipal waste-incineration plants.

9. Directive 94/67/EC on the incineration of hazardous waste.

10. Directive 92/112/EEC on procedures for harmonizing the programmes for the reduction and eventual elimination of pollution caused by waste from the titanium oxide industry.

11. Directive 88/609/EEC on the limitation of emissions of certain pollutants into the air from large combustion plants, as last amended by Directive 94/66/EC.

12. Directive 76/464/EEC on pollution caused by certain dangerous substances discharged into the aquatic environment of the Community.

13. Directive 75/442/EEC on waste, as amended by Directive 91/156/EEC.

14. Directive 75/439/EEC on the disposal of waste oils.

15. Directive 91/689/EEC on hazardous waste.

ANNEX III

INDICATIVE LIST OF THE MAIN POLLUTING SUBSTANCES TO BE TAKEN INTO ACCOUNT IF THEY ARE RELEVANT FOR FIXING EMISSION LIMIT VALUES

AIR

1. Sulphur dioxide and other sulphur compounds.
2. Oxides of nitrogen and other nitrogen compounds.
3. Carbon monoxide.
4. Volatile organic compounds.
5. Metals and their compounds.
6. Dust.
7. Asbestos (suspended particulates, fibres).
8. Chlorine and its compounds.
9. Fluorine and its compounds.
10. Arsenic and its compounds.
11. Cyanides.
12. Substances and preparations which have been proved to possess carcinogenic or mutagenic properties or properties which may affect reproduction via the air.
13. Polychlorinated dibenzodioxins and polychlorinated dibenzofurans.

WATER

1. Organohalogen compounds and substances which may form such compounds in the aquatic environment.
2. Organophosphorus compounds.
3. Organotin compounds.

4. Substances and preparations which have been proved to possess carcinogenic or mutagenic properties or properties which may affect reproduction in or via the aquatic environment.

5. Persistent hydrocarbons and persistent and bioaccumulable organic toxic substances.

6. Cyanides.

7. Metals and their compounds.

8. Arsenic and its compounds.

9. Biocides and plant health products.

10. Materials in suspension.

11. Substances which contribute to eutrophication (in particular, nitrates and phosphates).

12. Substances which have an unfavourable influence on the oxygen balance (and can be measured using parameters such as BOD, COD, etc.).

ANNEX IV

Considerations to be taken into account generally or in specific cases when determining best available techniques, as defined in Article 2(11), bearing in mind the likely costs and benefits of a measure and the principles of precaution and prevention:

1. the use of low-waste technology;

2. the use of less hazardous substances;

3. the furthering of recovery and recycling of substances generated and used in the process and of waste, where appropriate;

4. comparable processes, facilities or methods of operation which have been tried with success on an industrial scale;

5. technological advances and changes in scientific knowledge and understanding;

6. the nature, effects and volume of the emissions concerned;

7. the commissioning dates for new or existing installations;

8. the length of time needed to introduce the best available technique;

9. the consumption and nature of raw materials (including water) used in the process and their energy efficiency;

10. the need to prevent or reduce to a minimum the overall impact of the emissions on the environment and the risks to it;

11. the need to prevent accidents and to minimize the consequences for the environment;

12. the information published by the Commission pursuant to Article 16(2) or by international organizations.

Index

Printed in the United States
1535700002B/33

9 789041 197184